AGRICULTURE ISSUES AND POLICIES

MANURE MANAGEMENT

SELECT RESEARCH AND LEGAL ISSUES

AGRICULTURE ISSUES AND POLICIES

Additional books in this series can be found on Nova's website
under the Series tab.

Additional e-books in this series can be found on Nova's website
under the e-book tab.

AGRICULTURE ISSUES AND POLICIES

MANURE MANAGEMENT

SELECT RESEARCH AND LEGAL ISSUES

SHEILA E. CORBY
EDITOR

nova publishers
New York

For permission to use material from this book please contact us:
Telephone 631-231-7269; Fax 631-231-8175
Web Site: http://www.novapublishers.com

Library of Congress Cataloging-in-Publication Data

ISBN: 978-1-62948-874-5

Published by Nova Science Publishers, Inc. † New York

CONTENTS

PREFACE

In 2003, EPA introduced revised Clean Water Act regulations to better protect surface waters from nutrients from concentrated animal feeding operations (CAFOs). When applying manure to crop or pasture land (the primary disposal method), CAFOs now must follow a nutrient management plan that specifies a manure application rate that minimizes the threat to water quality. This book focuses on managing manure to improve air and water quality; trends and developments in hog manure management; and current laws and legislative issues with animal waste and hazardous substances.

Chapter 1 – Animal waste from confined animal feeding operations is a potential source of air and water quality degradation from evaporation of gases, runoff to surface water, and leaching to ground water. This chapter assesses the potential economic and environmental tradeoffs between water quality policies and air quality policies that require the animal agriculture sector to take potentially costly measures to abate pollution. A farm-level analysis of hog farms estimates the economic and environmental tradeoffs that occur when policies are designed to address pollutant flows to one environmental medium without considering flows to another medium. A national analysis addresses the broader impacts of coordinated (water and air) policies, including long-term structural adjustments and welfare impacts on both producers and consumers. The chapter also analyzes the potential implications of adding air quality regulations to existing Clean Water Act regulations in the Chesapeake Bay watershed, where a limited land base increases producers' costs of meeting manure management requirements.

Chapter 2 – In the past decade, hog production has increasingly become consolidated, with larger operations producing a greater volume of hog manure on smaller areas. With less crop-land for spreading the manure, hog farmers may be compensating through more effective manure management. The authors use data from 1998 to 2009 collected in three national surveys of hog farmers. Over this period, structural changes in the hog sector altered how manure is stored and handled. Changes to the Clean Water Act, State regulations, and local conflicts over air quality also affected manure management decisions. The findings further suggest that environmental policy has influenced conservation-compatible manure management practices. The authors examine how the use of nutrient management plans and of practices such as controlled manure application rates vary with scale of production and how these practices changed over the study period. This chapter is an update of an earlier report, *Changes in Manure Management in the Hog Sector: 1998-2004.*

Chapter 3 - The animal sector of agriculture has undergone major changes in the last several decades: organizational changes within the industry to enhance economic efficiency have resulted in larger confined production facilities that often are geographically concentrated. These changes, in turn, have given rise to concerns over the management of animal wastes and potential impacts on environmental quality.

Federal environmental law does not regulate all agricultural activities, but certain large animal feeding operations (AFOs) where animals are housed and raised in confinement are subject to regulation. The issue of applicability of these laws to livestock and poultry operations— especially the Comprehensive Environmental Response, Compensation, and Liability Act (CERCLA, the Superfund law) and the Emergency Planning and Community Right-to-Know Act (EPCRA)—has been controversial and has drawn congressional attention.

Both Superfund and EPCRA have reporting requirements that are triggered when specified quantities of certain substances are released to the environment. In addition, Superfund authorizes federal cleanup of releases of hazardous substances, pollutants, or contaminants and imposes strict liability for cleanup and injuries to natural resources from releases of hazardous substances.

Superfund and EPCRA include citizen suit provisions that have been used to sue poultry producers and swine operations for violations of those laws. In two cases, environmental advocates claimed that AFO operators had failed to report ammonia emissions, in violation of Superfund and EPCRA. In both cases, federal courts supported broad interpretation of key terms defining applicability of the laws' reporting requirements. Three other cases not dealing with reporting violations also have attracted attention, in part because of questions of whether animal wastes contain hazardous substances that can create cleanup and natural resource damage liability under Superfund.

In December 2008, EPA issued a rule to exempt animal waste emissions to the air from most CERCLA and EPCRA reporting requirements. Legal challenges to the rule followed. In October 2010, a federal court approved the government's request to remand the rule to EPA for reconsideration and possible modification, but the agency has not yet proposed a new or revised rule.

The lawsuits testing the applicability of CERCLA and EPCRA to poultry and livestock operations and potential changes by EPA to the 2008 exemption rule have led to congressional interest in these issues. In the 112th Congress, legislation was introduced that would amend CERCLA to clarify that manure is not a hazardous substance, pollutant, or contaminant under that act and that the notification requirements of both laws would not apply to releases of manure (H.R. 2997 and S. 1729). Proponents have argued that Congress did not intend that either of these laws apply to agriculture and that enforcement and regulatory mechanisms under other laws are adequate to address environmental releases from animal agriculture. Opponents respond that enacting an exemption would severely hamper the ability of government and citizens to know about and respond to releases of hazardous substances caused by an animal agriculture operation. Similar legislation has not been introduced in the 113th Congress.

In: Manure Management
Editor: Sheila E. Corby
ISBN: 978-1-62948-874-5

Chapter 1

MANAGING MANURE TO IMPROVE AIR AND WATER QUALITY*

Marcel Aillery, Noel Gollehon, Robert Johansson, Jonathan Kaplan, Nigel Key and Marc Ribaudo

ABSTRACT

Animal waste from confined animal feeding operations is a potential source of air and water quality degradation from evaporation of gases, runoff to surface water, and leaching to ground water. This chapter assesses the potential economic and environmental tradeoffs between water quality policies and air quality policies that require the animal agriculture sector to take potentially costly measures to abate pollution. A farm-level analysis of hog farms estimates the economic and environmental tradeoffs that occur when policies are designed to address pollutant flows to one environmental medium without considering flows to another medium. A national analysis addresses the broader impacts of coordinated (water and air) policies, including long-term structural adjustments and welfare impacts on both producers and consumers. The chapter also analyzes the potential implications of adding air quality regulations to existing Clean Water Act regulations in the Chesapeake Bay watershed, where a limited land base increases producers' costs of meeting manure management requirements.

* This is an edited, reformatted and augmented version of the United States Department of Agriculture Publication, ERR-9, dated September 2005.

Keywords: Animal waste, nitrogen, ammonia, water quality, nutrient management plan, manure management costs, price and quantity adjustments, CAFO

SUMMARY

U.S. environmental laws tend to focus on a single environmental medium (e.g., Clean Water Act, Clean Air Act, and Endangered Species Act). When a single pollution source can simultaneously affect more than one environmental medium, a single-medium approach to pollution control can confound policymakers concerned with economic efficiency. An uncoordinated set of policies that independently address different pollution issues can result in unnecessary and unanticipated economic and environmental costs.

Animal agriculture is facing just this situation. Animal agriculture produces a variety of pollutants, including organic matter, urea, ammonia, nitrous oxide, phosphorus, methane, carbon dioxide, pathogens, antibiotics, and hormones. Regulations to restrict animal farm emissions to the water might inadvertently increase emissions to the air, and vice versa.

What Is the Issue?

In 2003, EPA introduced revised Clean Water Act regulations to better protect surface waters from nutrients from concentrated animal feeding operations (CAFOs). When applying manure to crop or pasture land (the primary disposal method), CAFOs now must follow a nutrient management plan that specifies a manure application rate that minimizes the threat to water quality. The cost to farmers of meeting this requirement can be relatively high, primarily from moving manure to an adequate land base. A logical response by producers operating under a nitrogen-based plan might be to reduce the nitrogen content of manure spread on fields by enabling nitrogen to volatilize into the atmosphere from uncovered lagoons or by applying animal waste to land without incorporating it into the soil. But doing so also releases ammonia emissions into the air. As animal feeding operations are the primary source of ammonia in the United States, air quality regulations might require some States to regulate ammonia emissions from animal feeding operations.

What Did the Study Find?

Air and water quality regulations would be most cost effective if implemented simultaneously. This would allow farmers to select the most appropriate mix of practices that satisfy environmental quality goals while maximizing net returns. If environmental policies are uncoordinated, farmers may have to make costly changes to practices more than once before both environmental goals can be met.

To meet a water quality goal, farmers tend to use practices that increase ammonia emissions to the air. Similarly, the practices used to meet an air quality goal would tend to increase nitrogen losses from fields to ground and surface waters. Meeting both air and water quality goals would likely cost more than meeting either air or water goals.

Depending on how the air quality regulations are applied, this could have two impacts on CAFOs and water quality. First, farms identified as CAFOs might need to increase the amount of land on which they spread manure in order to continue to meet nutrient application standards. This could be particularly costly in a region where animal concentrations are high and cropland available for spreading manure is relatively scarce. For example, requiring CAFOs in the Chesapeake Bay watershed to control ammonia emissions would increase producer costs of land-applying manure by $4 million per year. Failure to account for these costs when developing an ammonia regulation could lead to the false conclusion that the policy is efficient.

Second, an uncoordinated approach could reduce water quality. If ammonia reductions were required of both CAFOs and smaller farms, the water quality benefits from the regulations restricting CAFOs' nitrogen emissions might be diluted by increased nutrient applications on the smaller farms, which have no such nitrogen restrictions. In the Chesapeake Bay watershed, for example, the nutrient content of manure produced on farms not covered by a nitrogen application standard would more than double if ammonia restrictions were applied to all animal feeding operations. This would increase the risk of nitrogen runoff into the Chesapeake Bay.

Anticipating the different forms and pathways that nitrogen takes can keep air quality and water quality policies from working at cross purposes. Then, true solutions—like diet manipulation (to reduce the amount of nitrogen excreted by animals) or industrial uses of manure—might become clearer.

How Was the Study Conducted?

The study used three separate but related analyses to capture the full range of economic decisions (and consequences) that result from farmers' meeting environmental regulations. Data from the 1998 Agricultural Resources Management Survey of hog producers were used to estimate the tradeoffs that occur at the farm level when policies are designed to address pollutant flows to one environmental medium without considering flows to another medium. The broader impacts of coordinated policies, including the welfare impacts on both producers and consumers and regional shifts in production, were examined with a national model of the agriculture sector that tracks nitrogen losses to the environment. A case study of the Chesapeake Bay watershed was used to demonstrate the problems that hypothetical ammonia reductions would have for farms meeting the CAFO regulations in a region where land for spreading manure is relatively scarce, and for resource managers trying to reduce nitrogen loads.

At the heart of all three analyses are nitrogen loss coefficients that are derived from a mass-balance accounting of nitrogen in manure. These were obtained from EPA, and enabled us to estimate tradeoffs in nitrogen losses to the air and nitrogen applied to land as different manure management practices are employed.

SECTION 1. INTRODUCTION

U.S. agriculture produces affordable food and fiber for domestic use and export and contributes significantly to the economic base of rural communities. But many agricultural activities also produce pollutants that can harm the environment. For example, animal production generates byproducts such as organic matter, urea, ammonia, nitrous oxide, phosphorus, methane, carbon dioxide, pathogens, antibiotics, and hormones. Without proper management, these materials can degrade surface water, ground water, and soils. Environmental policy aims to improve the management of agricultural systems such that environmental harm is minimized.

Mitigating pollution problems can challenge policymakers when more than one environmental medium is affected by a single pollution source. The correction of a single problem without simultaneously addressing others may not increase societal welfare as much as anticipated, and may even decrease it. Thus, an uncoordinated set of policies that independently address different pollution issues could result in unnecessary losses in societal welfare. Scientists and program managers are also aware of these tradeoffs. However, environmental laws often focus on only one environmental medium (Clean Water Act, Clean Air Act, Endangered Species Act). Such narrowly focused programs may harbor large opportunity costs, especially with high interdependence in pollution flows between different environmental media (U.S. EPA, 1996).

Animal agriculture, in particular, has faced increasing environmental regulations in recent years. Growth and concentration in the industry over the past several decades has prompted concerns over environmental degradation in areas where production facilities are clustered. Concentrated animal feeding operations (CAFOs) have been regulated since 1974 under the Clean Water Act. CAFO regulations were strengthened in 2003 to reduce the threat of nutrients entering surface water, and were the subject of an earlier ERS study (Ribaudo et al., 2003). But these regulations do not require control of potential air emissions from CAFOs. Confined animal operations are the largest source of ammonia emissions in the United States (Abt Associates, 2000). Ammonia emissions have long been encouraged as a justifiable byproduct of meeting water quality goals (Sweeten et al., 2000). Lagoons, for instance, are commonly used to store and treat manure waste from swine operations. These storage systems volatilize nitrogen, thereby reducing its concentration in lagoon effluent and reducing the cost of meeting land application requirements. But, the volatilized nitrogen compounds escape into the air, creating odors, contributing to fine particulates (haze), and hastening global climate change (National Research Council, 2003). Only recently has ammonia loss been viewed as a potential problem in terms of air quality (Sweeten et al., 2000).

The current uncoordinated approach to air and water quality protection has potentially costly implications for both animal producers and society in general. Some animal feeding operations already subject to water quality regulations may soon be required to meet ammonia emission regulations. Technologies adopted to reduce water pollution may be inadequate for meeting both water quality and ammonia requirements, and might have to be abandoned or modified, at some cost, to comply with both sets of regulations.

Smaller operations not required to meet Clean Water Act regulations might be required to meet air quality regulations. If they change manure management practices to reduce ammonia emissions, the nitrogen content of the operations' manure will increase. If manure applications to the land remain unchanged, the risk of nitrogen runoff to water resources increases. A more coordinated approach to environmental quality protection could avoid these unintended consequences.

Research Objectives

This chapter assesses the potential economic and environmental tradeoffs between air and water quality when the animal sector is required to take potentially costly measures to abate pollution. To date, only a few analyses have discussed the cross-media problem (Helfand, 1994; Hohmann, 1994; Resources for the Future, 1996), and none explore the theoretical or empirical tradeoffs inherent in cross-media environmental policy. We extend this literature by acknowledging that multiple pollutants from animal feeding operations may enter different media; pollution control technologies effective in one environmental medium may conflict with technologies to control pollutants to other media. Examining the implications of regulating across environmental media may help guide future air and water quality regulations and improve the performance of existing policies.

To accomplish these objectives, this chapter:

- Reviews some of the potential pollution problems attributable to animal waste, the physical relationships inherent in the waste stream that complicate efficient manure management policy, and the environmental policy regime facing animal agriculture.
- Estimates the tradeoffs that occur at the farm level for hogs when policies are designed to address pollutant flows to one environmental medium without considering flows to another medium. This analysis best captures the production decisions that an individual producer makes when faced with market signals and regulatory requirements in the context of the farm's capital and resource bases.
- Analyzes the national impacts of coordinated policies, including the welfare impacts on both producers and consumers. This accounts for the price effects and regional adjustments missing from the farm-level analysis.
- Analyzes the implications of adding air quality regulations to existing Clean Water Act (CWA) regulations in a region where a limited land base increases the costs of meeting manure management requirements. This case study of the Chesapeake Bay watershed demonstrates how the costs of meeting CWA requirements are

affected if ammonia emissions must also be reduced. It also demonstrates how water quality might be affected if ammonia reductions are required on farms not covered under the CWA.

SECTION 2. ANIMAL AGRICULTURE AND THE ENVIRONMENT

The U.S. animal sector has undergone major structural changes over the past several decades, the result of domestic and export market forces, technological changes, and industry adaptations. The number of large confined production units has expanded, while animal production and feed production are increasingly separated. The total number of animal units increased by about 10 percent between 1987 and 1997, while the number of animal feeding operations (AFOs) decreased by more than half (Gollehon et al., 2001).

Growing concerns about the potential impacts of these changes on environmental quality have spurred local, State, and Federal action. Complaints about water quality and air quality (primarily odor) fuel most of the conflicts between the animal sector and the general population. The Environmental Protection Agency (U.S. EPA) revised Clean Water Act regulations in 2003 for controlling runoff of manure nutrients from the largest AFOs. At the State level, North Carolina entered a legal agreement with the State's largest swine producers to develop innovative waste management strategies that would replace uncovered lagoon and spray-field systems as a means of storing and treating waste from large hog operations, in order to prevent a repeat of the massive damage to water resources caused by Hurricane Floyd in 1995 (Williams, 2004). Rules on the handling of poultry litter now protect water supplies in eastern Oklahoma, which has witnessed a sharp increase in large poultry operations (Cody, 2003). The South Coast Air Quality Management District in California has introduced new rules for handling and disposing of dairy manure in order to reduce ammonia emissions that have affected heavily populated areas downwind (Wilson, 2004). Iowa, Pennsylvania, Arkansas, and Kentucky are among States that have introduced rules for curbing water pollution, ammonia, and odor from AFOs (Patton and Seidl, 1999; U.S. EPA, 2002).

Environmental Impacts of Animal Production

The major source of environmental degradation from AFOs is waste products (manure, urine, and bedding material). Pollution from animal waste includes runoff of nutrients, organic matter, and pathogens to surface water, leaching of nitrogen and pathogens to ground water, volatilization of gases and odors to the atmosphere, and emissions of fine particulates. Pollutants can originate at several stages of production, including:

- Production houses where animals are confined;
- Manure storage structures such as tanks, ponds, and lagoons;
- Land where manure is applied.

The focus of this chapter is primarily on the various forms of nitrogen generated by manure. Nitrogen moves freely between the soil, air, and water, and there is a high degree of interdependence between the forms and paths it takes. Nitrogen from manure is therefore capable of affecting the quality of more than one environmental medium.

Nitrogen Products from Animal Production

Nitrogen is found throughout the environment. It is required by all living things and is a critical crop nutrient. Seventy-eight percent of the atmosphere consists of elemental nitrogen gas (N_2), which is inert, does not affect environmental quality, and is unavailable to living organisms. However, during the past 50 years, large amounts of reactive N have been added to terrestrial systems—including ammonia, ammonium, nitric oxide, nitrogen dioxide, nitrate, and nitrite—through combustion of fossil fuels and production/application of synthetic and organic fertilizer (Follett and Hatfield, 2001). Reactive forms of nitrogen pose a potential threat to environmental quality because of their ability to combine with other compounds and create environmental problems (see box, "Ammonia and Nitrate in the Environment," p. 6). Emissions of reactive nitrogen from AFOs are substantial, and their control represents an important objective in improving environmental quality (NRC, 2003).

Water-Air Interactions

Emissions of nitrogen to water and to the atmosphere are not independent events, but are linked by the biological and chemical processes that produce the various nitrogen compounds (figure 2-1). Nitrogen enters the system in animal feed. Some of the nitrogen is retained in the animal products (meat, milk, eggs), but as much as 95 percent is excreted in urine and manure (Follett and Hatfield, 2001).

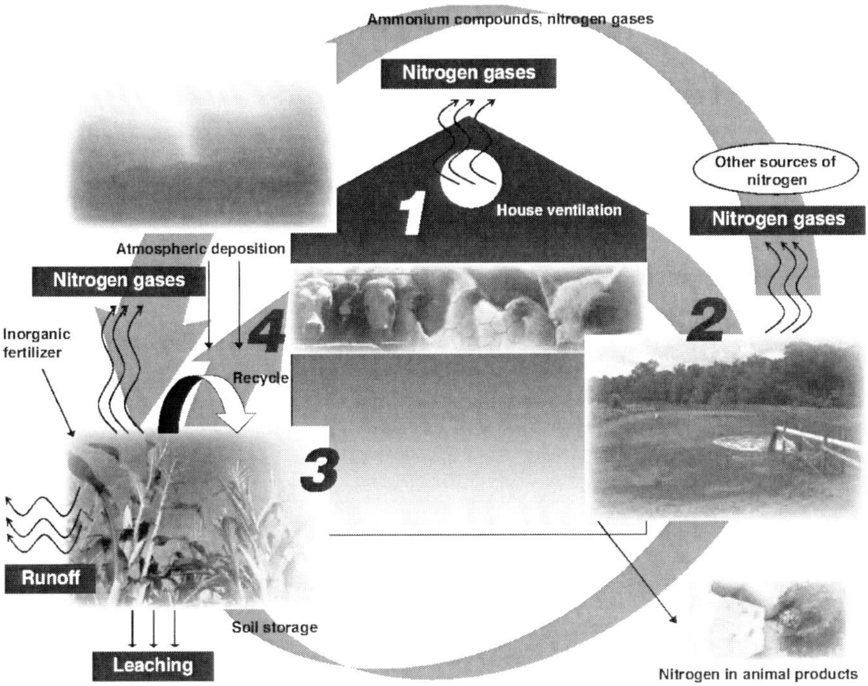

The nitrogen cycle is a complex one, without a beginning, middle, or end. The principle of mass-balance ensures that the amount of nitrogen in a closed system is constant. Thus, any action to divert it from one pathway must necessarily transfer it into another. In this stylized figure:

1. Animals in the "house" release nitrogen in three ways: they produce manure (which then enters a storage system); they store nitrogen internally, which is bound in animal products distributed to markets; and they produce gases (directly and indirectly in manure production), which are released as air emissions;

2. Manure is stored in lagoons, tanks, pits, or other structures before being transported to fields for use as fertilizer;

3. Manure nitrogen applied to fields may be stored in the soil, leached into groundwater, run off into surface water, volatilize into air emissions, and be bound in crops; or

4. Nitrogen bound in crops may be used for feed for the animals, and the cycle begins again.

Nitrogen also enters and exits the system through intermediate pathways. For example, some of the nitrogen released into the air will settle back on the fields (deposition) and some new nitrogen will be added in the form of commercial fertilizer and nitrogen deposition from other sources.

Figure 2-1. Nitrogen follows many pathways in a livestock operation.

Manure can collect in or under the production house for a few hours or several years, depending on the collection system. Production houses are ventilated to expel gases that are emitted, including ammonia. The manure is eventually removed from the house to a storage structure (lagoon, tank, pit, or slab) and stored anywhere from a few days to many months. Losses of nitrogen to air and water can occur during this time, depending on the system and the extent of contact with rain and wind. The stored manure is eventually transported to fields where it is applied. Losses to air and water from the field vary, depending on application method and rate. Nitrogen in the field helps produce crops, which may in turn be fed to animals, thus completing the cycle. Nitrogen lost to the air eventually returns to earth, where it can be a source of plant nutrients, or contribute to runoff or leaching to water resources.

Ammonia and Nitrate in the Environment

Two nitrogen compounds of particular environmental concern are ammonia and nitrate. Ammonia is a pungent gas that is a potential health hazard to humans and livestock. Nitrogen in animal manure can be converted to ammonia by a combination of hydrolysis, mineralization, and volatilization (Oenema et al., 2001). Once in the atmosphere, ammonia can be converted rapidly to ammonium aerosol by reactions with acidic compounds such as nitric acid or sulfuric acid. As an aerosol, ammonium contributes directly to fine particulates, the source of haze in the atmosphere. Gaseous ammonia that is not converted to ammonium is removed from the atmosphere by dry deposition, while ammonium aerosol is primarily removed from the atmosphere by rain or snow. Atmospheric deposition can cause ecological problems by changing plant communities through nutrient enrichment and soil acidification. Large quantities of ammonia are emitted from animal operations each year, making up 50 to 70 percent of annual ammonia emissions from all sources in the United States (NRC, 2003).

Nitrate is a highly soluble compound that is an important plant nutrient. In water, it can degrade water quality by spurring eutrophication. Nitrate can be a human or animal health hazard in drinking water if in high enough concentrations. The U.S. EPA has established a maximum contaminant level for drinking water of 10 ppm for nitrate-nitrogen. Nitrate enters water resources through runoff or leaching from fields receiving manure, or from leaks in manure storage structures.

The form nitrogen takes in its journey from animal to field depends on a host of factors, including storage technology, manure moisture content, temperature, air flow, pH, and the presence of micro-organisms. Reducing nitrogen movement along one path by changing its form will increase nitrogen movement along a different path (NRC, 2003). For example, reducing ammonia losses from a field to the atmosphere by injecting waste directly into the soil increases the amount of nitrogen at risk of moving to water resources as nitrate (Oenema, 2001; Abt Associates, 2000). Ignoring the interactions of the nitrogen cycle in developing manure management policies could lead to unintended and adverse effects on environmental quality.

Economic Relationships in Animal Waste Management

Farmers consider an array of factors when deciding about farm management and conservation practices. Production decisions are based on market prices, the farm's resources, available technologies, management skill, and expectations about weather. But incentives to consider environmental quality in the balance are often lacking.

How an animal feeding operation manages waste determines the paths of nitrogen and other manure constituents from production to disposal. Many practices are available for reducing gaseous emissions, runoff, and leaching (see box, "Manure Management Strategies"), but manure management practices with strictly public benefits (benefits realized off the farm) will be little used unless economic and regulatory conditions change.

Environmental policy incentives can be subsidies that favor a set of practices (technology incentives), requirements that certain practices be used (technology standards), or requirements that particular farm-level environmental goals be achieved (performance standards) (Ribaudo, Horan, and Smith, 1999). In each case, the farmer will choose that set of practices that maximizes net returns while taking into account the financial incentives or constraints introduced by policy. Whether environmental goals are subsequently met depends on how well the incentives are designed by the resource agency.

Policy Regime for Animal Waste

The major Federal environmental law currently affecting animal feeding operations is the Clean Water Act. AFOs concentrate animals, feed, manure, and urine in a small land area. Feed is brought to the animals rather than their grazing in pastures, fields, or on rangeland. The Clean Water Act specifies that AFOs may be covered by the National Pollutant Discharge Elimination System (NPDES) program (U.S. EPA, 2003). NPDES permits are required by point sources (facilities that discharge directly to water resources through a discrete ditch or pipe) before they can discharge into navigable waters. The permits specify a level of treatment for each effluent source. The regulations set thresholds for size categories based on the number of animals confined at the operation for a total of 45 days or more in any 12-month period.

MANURE MANAGEMENT STRATEGIES

Diet and health—Feed additives and improved nutrient utilization in animals' diets can simplify manure management at all stages of handling and disposal by reducing the amount of nutrients in waste (CAST, 2002; Abt, 2000). The goals of feed management are to match nutrient needs of animals more closely with nutrients in feed. Animal genetics, phase feeding (altering feed to match the age or production level of animals), and amino acid supplements are management tools that are currently available to animal producers.

Chemical addition—Chemicals can be added to manure during its collection in order to bind odorous compounds and to reduce ammonia emissions by lowering pH. For example, field tests indicate that alum can reduce ammonia emissions by 75-97 percent when added to poultry litter (Moore et al., 2000). Alum also increases the nitrogen content of litter that is eventually spread on fields, potentially increasing ammonia emissions from fields and loss of nitrate to water resources.

Air treatment—Trapping air vented from production houses and treating it before discharge to the atmosphere can reduce odorous compounds, ammonia, and other gases. Treatment processes include ozonation and biofilters (Jacobson et al., 1999). These processes do not affect the nitrogen content of manure.

Solid-liquid separation—Separating urea from solid fecal matter either mechanically or with sedimentation basins avoids some of the reactions that cause the formation of ammonia and odor. Separation also simplifies waste handling and disposal (Jacobson et al., 1999).

While reducing ammonia emissions, separation preserves nitrogen in liquid and solid wastes, potentially increasing losses to air and water when eventually applied to fields.

Tank covers—Covering storage tanks can greatly reduce the discharge of ammonia to the atmosphere, primarily by altering pH and preventing the formation of ammonia (Jacobson et al., 1999). Storage tanks can be covered with a roof, concrete lid, or flexible plastic cover. The surface of the stored waste can also be covered with straw or other materials (polystyrene foam, air-filled clay balls). This material serves as a medium for micro-organisms that act as a biofilter. While reducing ammonia emissions, covers also increase the nitrogen content of effluent that is eventually spread on fields, increasing the potential for both ammonia emissions and loss of nitrate to water resources.

Lagoon covers—Plastic covers that float on the lagoon surface or that are tented over lagoons can greatly decrease gaseous emissions (Jacobson et. al.,1999, Arogo et al., 2002). Some systems (anaerobic digesters) also capture methane and use it as a biofuel to generate electricity. Covering a lagoon prevents the formation of ammonia by lowering the pH, but increases the nitrogen content of the effluent that is eventually sprayed on fields. While ammonia emissions from fields sprayed with lagoon effluent might increase, the net effect is a reduction in ammonia emissions from both lagoon storage and field applications. However, the risk of nitrate loss to water increases.

Manure incorporation and injection—Rapidly incorporating manure into the soil, either by plowing or disking solids after spreading or injecting liquids and slurries directly into the soil, reduces odor and ammonia emissions (Abt, 2000; Arogo et al., 2002). But, this also increases the nitrogen available for crops in the soil, and thus the risk of nitrate runoff to water resources.

Comprehensive nutrient management—Following a comprehensive nutrient management plan when applying manure and commercial fertilizer to land can reduce potential losses of nitrate to water resources through runoff or leaching (USDA, NRCS, 2005). Nutrient management matches nutrient applications to crop needs so that as few nutrients as possible are lost to the environment.

Individual components include testing manure and the soil for nutrient content, calibrating application equipment, balancing crop needs with commercial fertilizer, and recordkeeping. Plans can account for atmospheric losses of nitrogen from animal operations, as well as atmospheric deposition of nitrogen on cropland.

Concentrated animal feeding operations (CAFOs) are AFOs that are designated by the U.S. EPA or State authority as requiring a point-source discharge permit. We estimate the number of potential CAFOs by using the December 2004 EPA specifications of CAFOs (see box, "EPA Size Thresholds") based on operation size. Large operations are considered CAFOs. Medium-sized operations *can* be designated CAFOs if a manmade ditch, pipe, or similar device carries manure or wastewater from the operation to surface water or the animals come into contact with surface water that runs through the confinement area, or the operation is designated by the permitting authority (typically the State). Even a small operation can be designated a CAFO by the permitting authority if it is found to be a significant source of pollution and it meets the above conditions.

EPA size thresholds

Animal sector	Size threshold (number of animals)		
	Large CAFOs	**Medium CAFOs**	**Small CAFOs**
Cattle or cow/calf pairs	1,000 or more	300 - 999	Fewer than 300
Mature dairy cattle	700 or more	200 - 699	Fewer than 200
Veal calves	1,000 or more	300 - 999	Fewer than 300
Swine (over 55 pounds)	2,500 or more	750 - 2,499	Fewer than 750
Swine (under 55 pounds)	10,000 or more	3,000 - 9,999	Fewer than 3,000
Turkeys	55,000 or more	16,500 - 54,999	Fewer than 16,500
Laying hens or broilers (liquid manure handling systems)	30,000 or more	9,000 - 29,999	Fewer than 9,000
Chickens other than laying hens (other than a liquid manure handling system)	125,000 or more	37,500 - 124,999	Fewer than 37,500
Laying hens (other than a liquid manure handling system)	82,000	25,000 - 81,999	Fewer than 25,000

Source: U.S. Environmental Protection Agency.

The NPDES permits for CAFOs contain technology-based effluent limit guidelines for the production area and for the land receiving manure. A CAFO must have a nutrient management plan, covering the land receiving manure, that specifies an application rate for manure nitrogen or phosphorus based on the agronomic needs of the crops. EPA estimated that about 15,500 AFOs would be defined or designated as CAFOs (U.S. EPA, 2003).[1] This represents about 5 percent of U.S. AFOs, but covers about 60 percent of all confined animals.

Atmospheric emissions of pollutants are regulated by the Clean Air Act (CAA). The CAA authorizes regulatory programs primarily for protecting human health. EPA has recently initiated development of regulations for reducing fine particulates in the atmosphere (referred to as PM2.5, for particles less than 2.5 microns in size). The Clean Air Act requires State, local, and tribal governments to identify areas not meeting national air quality standards for fine particles (one of the six criteria pollutants regulated under the Act) (U.S. EPA, 2005). States with designated non-attainment areas must submit plans by February 2008 that outline how they will meet the standards by 2010. This regulation could affect animal operations because ammonia is a major precursor of fine particulates. Controlling ammonia from animal operations would be a likely priority in non-attainment areas with high concentrations of animals (U.S. EPA, 2000).

Also covering air pollution is the Comprehensive Environmental Response, Compensation and Liability Act (CERCLA), enacted in 1980 to provide broad Federal authority to respond to releases of hazardous substances that might endanger public health. Section 103 of CERCLA requires that the person in charge of a facility immediately notify the National Response Center as soon as he has knowledge of any release in quantities equal to or greater than the "reportable quantity" (e.g., for ammonia, 100 pounds in a 24-hour period) of a hazardous substance. EPA is authorized to require remedial action when appropriate. Although CERCLA is focused on hazardous wastes from industrial plants, the increased size and consolidation of animal feeding operations may make their ammonia and hydrogen sulfide emissions subject to the notification provisions (U.S. EPA, 2005).

Neither the Clean Air Act nor CERCLA currently recognize CAFOs for regulatory purposes. There is no reason to expect that either of these laws would be applied to the same set of operations required by the Clean Water Act to meet nutrient application standards.

While environmental policies may not explicitly recognize the compound (air and water) effects of manure management practices, these interactions are well known to soil and animal scientists. NRCS specifically recognizes the potential air quality impacts of manure management practices that could be included in a Comprehensive Nutrient Management Plan (CNMP). A CNMP is a conservation plan unique to AFOs that addresses natural resource concerns dealing with soil erosion, manure, and organic byproducts and their potential impacts on water quality (USDA, NRCS, 2005). A CNMP must meet strict technical standards. NRCS recommends that AFOs consider the impact of selected conservation practices on air quality during the CNMP development process. However, NRCS does not currently maintain technical standards for practices that address air quality concerns, so they are not included in a CNMP. In addition, CNMPs are not required of AFOs under EPA regulations. The nutrient management plan that permitted CAFOs must implement is not the same as a CNMP.

Multi-Path Emissions Have Unique Economic Implications

Pollution imposes costs on society that are not borne by the polluter. Thus, pollution control policies can increase societal welfare. Literature on the design of pollution control policies has almost exclusively focused on the control of one specific pollutant (Baumol and Oates, 1988; Tietenberg, 1990). The assumption that there is no interaction between pollutants, either in production or once emitted, is a strong one. Practical experience shows that this is not often the case.

When a production activity creates multiple pollutants or pollutes more than one environmental medium, correcting a single problem can lead to further resource misallocations (Lipsey and Lancaster, 1956). For example, Weinberg and Kling (1996) showed that addressing two pollution externalities associated with irrigation policy can increase welfare, even when the policies are introduced independently. However, there are situations where acting independently can lead to lower welfare than if one pollution problem was not addressed at all. Coordination of policies would ensure that an optimal mix of pollution control is achieved. Coordination is particularly important when policies are in conflict. When policies are complementary, addressing one pollutant tends to move the other in the desired direction. When policies are in conflict, addressing only one could worsen the other, even to the point of reducing overall economic welfare. Reducing soil erosion and improving water quality is an example of policies that can be in conflict (Conner et al., 1995; Lakshminarayan et al., 1995).

The multi-pollutant, multi-effect nature of air pollution has led to serious consideration of coordinated policies. For example, concern over the emission of acidic sulfur and nitrogen compounds in Europe led to separate protocols for reducing emissions of sulfur dioxide, nitrogen oxides, ammonia, and volatile organic compounds. Because of the synergistic effects of these pollutants in the atmosphere, and the multiple impacts on environmental quality (soil and water acidification, eutrophication, ozone formation), a combined approach offers the best means of achieving costeffective control (Metcalfe et al., 1998; Michaelis, 1992).

One consequence of an uncoordinated approach is uncertainty about future regulations and how they might affect production. Uncertainty complicates a firm's investment decisions, in light of the significant adjustment costs and irreversibility of investment that capital expenditures generally entail (Lee and Alm, 2004).

Policies in conflict would be particularly prone to this. If technologies and management measures implemented by CAFOs to meet water pollution regulations turn out to be in conflict with the goal of reducing air emissions, some of these measures might have to be abandoned, at significant cost to producers.

The compound and multi-pollutant nature of nitrogen flows in the environment suggests the importance of coordinated, rather than piecemeal, control (Bull and Sutton, 1998; Baker et al., 2001). In our chosen case of nitrogen from manure, CWA regulations require CAFOs to meet a nutrient standard for land application.

This generally means more land is required for spreading manure than has been used in the past, an expensive proposition for many large farms (Ribaudo et al., 2003). One rational management response under a nitrogen standard is to encourage volatilization of manure nitrogen (e.g., use of uncovered lagoons, surface application to fields) to reduce the nitrogen content in waste, thus allowing higher application rates on cropland and reducing the amount of land needed for spreading (Sweeten et al., 2000). But, such a strategy would increase atmospheric emissions of ammonia and worsen air quality. Environmental policies that simultaneously consider all the environmental consequences of manure nitrogen may increase social welfare relative to single-medium policies when pollutant emissions are interdependent. Zilberman et al. (2001) cite the multi-path nature of animal waste as one reason why current policies are inadequate. A policy focused on nitrogen applications to land allows the buildup of other potential pollutants in the soil, such as phosphorus, and ignores problems such as odor and dust.

Ammonia Coefficients Used in This Chapter

While there are growing concerns over ammonia emissions to the atmosphere, the science behind developing a national ammonia emissions inventory and assessing ammonia management practices is lacking (NRC, 2003). According to the NRC, "There is a general paucity of credible scientific information on the effects of mitigation technology on concentrations, rates, and fates of air emissions from AFOs, (p. 5)." Much of this has to do with lack of appropriate monitoring technologies and procedures for measuring emissions from a nonpoint source.

However, we are not trying to estimate national emissions, but to assess tradeoffs in the nitrogen cycle (figure 2-1). Recent research on manure management has provided useful information on the nature of these tradeoffs. We used as our starting point the manure management "trains" (MMT) developed by EPA (2004). This inventory of currently used animal production and manure management systems takes a mass balance or systems approach that is central to our study. However, these MMTs did not include management practices for reducing ammonia emissions. We adapted MMTs for systems incorporating recognized ammonia reduction technologies by using reduction efficiencies reported in the published scientific literature to redirect nitrogen along the appropriate paths. Table 2-1 shows the ammonia losses and the nitrogen available for crops for a set of common production and manure handling systems (with and without ammonia-reducing practices) that are used in this chapter to assess tradeoffs. The ammonia-reduction efficiency of lagoon covers (Jacobson et al., 1999; Arogo et al., 2002.; Oenema et al., 2001), incorporation (Moore and Meisinger, 2003; Hatfield, 2003; Jacobson et al.,1999; Abt, 2000) and alum (Moore et al., 2000) were obtained from the literature.

We make several assumptions to simplify our analysis.. We assume all CAFOs regulated under the Clean Water Act must adopt nitrogen-based nutrient management plans. An unknown number will probably have to adopt a phosphorus-based plan instead. Under such a plan, manure nitrogen is generally applied at rates less than crop needs, so supplemental commercial nitrogen fertilizer is required (Ribaudo et al., 2003). Incentives to encourage nitrogen losses to the atmosphere would not exist in this case. Subsequently reducing ammonia emissions would not pose increased risks to water quality as long as compensating adjustments in supplemental commercial nitrogen fertilizer are made and the increased nitrogen content of manure does not exceed crop needs.

Table 2-1. Manure management systems and nitrogen losses

Animal	System	N excreted	Losses from building	Losses from storage	Losses from field	Total losses to air	Total available for crops
Pounds N/head/year							
Hogs	Lagoon, uncovered	18.3	4.9	9.5	0.8	15.2	3.1
Hogs	Lagoon, covered	18.3	4.9	0.5	2.8	8.2	10.1
Hogs	Deep pit, surface apply	18.3	6	0	2.6	8.6	9.7
Hogs	Deep pit, incorporate	18.3	6	0	0.4	6.4	11.9
Dairy	Flush barn, surface apply	220	44	125	11.2	180.2	39.8
Dairy	Flush barn, incorporate	220	44	125	2.8	171.8	48.2
Dairy	Daily spread, surface apply	220	15.2	2.2	37.7	55.1	164.9
Dairy	Daily spread, incorporate	220	15.2	2.2	8.3	25.7	194.3
Poultry	Surface apply	0.9	0.18	0.03	0.17	0.38	0.51
Poultry	Incorporate	0.9	0.18	0.03	0.04	0.25	0.65
Poultry	Alum, surface apply	0.9	0.03	0.04	0.21	0.28	0.62
Fed beef	Solid storage, surface apply	102	0	20.8	13.8	34.6	67.4
Fed beef	Solid storage, incorporate	102	0	20.8	0.7	21.5	80.5

Source: U.S. EPA; Jacobson et al.; Arogo et al.; Oenema et al.; Moore and Meisinger; Hatfield; Jacobson et al.; Abt; Moore et al.

The Clean Air Act and CERCLA are national laws, but whether they affect animal operations depends on local conditions (non-attainment area in the Clean Air Act) or characteristics of individual operations (daily ammonia emissions under CERCLA). While these laws currently have only limited impacts on animal feeding operations, we do not attempt to forecast which regions or types of operations might be required to reduce ammonia emissions. Our goal is to demonstrate potential tradeoffs that could be important to farmers and to resource managers wherever animal feeding operations are required to meet water quality and/or air quality goals.

We also do not consider the fate of atmospheric nitrogen in our analyses. Atmospheric nitrogen can travel long distances, and our models cannot track its movement. Industrial and mobile sources also contribute significant amounts of atmospheric nitrogen. Accounting for atmospheric deposition would likely reduce the manure nitrogen application rates under a nutrient management plan used in our analyses, but would not alter our basic findings.

SECTION 3. WATER-AIR TRADEOFFS AT THE FARM LEVEL

A farm-level perspective allows us to look at the economic decisions individual producers make to meet environmental goals, given previous management choices and farm characteristics. We examine the hog sector because it exemplifies the changes to scale, structure, and location that have occurred in the confined animal sector since the 1960s. In 1982, there were 175,284 farms with confined hogs, containing 6.3 million animals (USDA, ERS, 2005). By 1997, the number of farms had shrunk 64 percent to 63,723, while the number of hogs on these farms increased to 8.2 million (USDA, ERS, 2005). These larger facilities (in terms of animal units) are not necessarily larger in terms of cropland. Thus, 51 percent of the recoverable nitrogen (nitrogen remaining after manure handling and storage) in confined hog manure in 1997 was estimated to be in excess of crop needs at the farm level (Gollehon et al., 2001).

Most confined hog operations use either a slurry pit system or a liquid lagoon system for managing manure. Slurry systems store undiluted, untreated manure in watertight tanks or pits until it can be land applied. Storage can be either under the house or outdoors. The stored slurry is surface applied to fields by sprayer trucks or wagons, or incorporated into the soil with chisel plows behind nurse tanks, or directly injected into the soil with drag hoses. Most ammonia emissions from these systems are from the field where manure is applied (U.S. EPA, 2004).

Lagoon systems use open holding ponds to treat diluted manure for an extended period of time. Lagoons stabilize organic matter, reduce the nutrient mass that must be land applied, and vent a large quantity of the manure nitrogen as ammonia. Some of the diluted lagoon liquid is used to flush the production houses. The "digested" lagoon liquid is eventually sprayed on cropland. Lagoons are used primarily in warmer climates where the anaerobic processes can take place year round. Lagoon systems emit more ammonia per animal unit to the atmosphere than do slurry systems (U.S. EPA, 2004).

Because of the high cost of transporting manure relative to the value of the nutrients in the manure, farmers have an incentive to overapply manure to land located near their livestock facilities. The amount of manure generated on CAFOs and its estimated nutrient content indicate that 82 percent of hog CAFOs were overapplying manure nitrogen in 1998 (Ribaudo et al., 2003). Farmers can reduce threats to water quality by testing soil and manure for nutrient content, and applying nutrients at rates consistent with the agronomic needs of crops. Such an approach could force farmers to spread manure on more land, often requiring manure to be transported greater distances from the hog facility (Ribaudo et al., 2003). Farmers faced with nitrogen application restrictions through a required nutrient management plan—but not ammonia emission restrictions—might try to reduce the nitrogen content of manure as a means of reducing the amount of land needed for spreading, and limiting hauling costs. The nitrogen content of manure can be reduced by promoting the creation of ammonia and its volatilization to the atmosphere. This can be done by storing manure in uncovered lagoons and by surface applying slurry rather than injecting it (Sweeten et al., 2000). For example, the nitrogen available to crops in lagoon liquid is 70 percent lower when coming from an uncovered lagoon rather than a covered lagoon.

If, on the other hand, farmers face restrictions in ammonia emissions but not runoff, they can reduce emissions by adopting storage structures and management methods that reduce manure's contact with the air and maintain a low pH. Preventing the formation of ammonia preserves the nitrogen content of manure, increasing the availability of nitrogen for crops, as well as the risk of nitrogen loss to surrounding waters if the land base receiving lagoon liquid stays the same.

To examine the effect of potentially conflicting policies on a farmer's production decisions, we constructed a hog farm economic model. We evaluate three scenarios: (1) a nitrogen application standard as part of a nutrient management plan required by the 2003 CAFO regulations under the Clean Water Act, (2) a hypothetical ammonia emission standard based on available emission abatement technologies, and (3) a coordinated policy that meets both land application and ammonia emission standards. A positive mathematical programming model with calibrated cost functions captures the essential farm-level tradeoffs between air emissions and water discharges of nitrogen (see Appendix A—web only—for details). Farmers maximize profits given input prices, output prices (hogs and crops), regulatory requirements, and available cropland by choosing a manure management technology, the amount of land on which to spread manure, the acreage of each crop to plant, the amount of commercial fertilizer to purchase, and the number of hogs to produce. We assumed that the basic manure storage system (pit or slurry) would not change.

In the model, nitrogen enters through the feed ration and is retained by the animals or excreted in manure. Once excreted, the nitrogen may be released into the atmosphere through ammonia emissions or preserved in the manure storage and handling system until it is applied to cropland. Nitrogen enters cropland through commercial and manure fertilizer applications. The crop retains some nitrogen, some is bound in the soil substrate, and some is released directly into the environment through air emission and water runoff. Water quality impacts are assumed to be directly related to the amount of nitrogen applied to cropland that is in excess of crop needs, after losses to the atmosphere. Air emissions are derived from total animal production and the type of storage/handling technology employed by the animal feeding operation.

The model is calibrated with data from the 1998 USDA-ARMS survey of hog operations, the most recent survey for hogs. In the analysis, eight representative CAFOs are depicted, corresponding to four major hog producing regions (East Corn Belt, West Corn Belt, Mid-Atlantic, and South and West) and two major manure storage technologies (lagoon and pit). We consider two technological options currently available to hog farms that influence the level of ammonia released to the air: the injection of manure into the soil and covering lagoons.

Single-Medium Environmental Policies

Baseline

Model results indicate how the two single-medium policies and a joint multimedia policy would alter farmers' decisions affecting production, input levels, nitrogen to soil and air, and the use of emission technologies relative to the baseline year 1998—the year of the survey to which the model is calibrated. Baseline costs and profits reflect production decisions made in the absence of any regulatory constraints. In the baseline year, all hog manure was applied onfarm to corn, soybean, and other crops at a rate 7.3 times greater than the nitrogen-based agronomic rate (table 3-1, column 1). This rate reflects the quantity of manure produced by farms relative to the amount of land on which manure was spread in 1998 and the crops reported as receiving manure. In the baseline year, about 10 times more ammonia nitrogen is released from manure storage facilities (lagoons and pits) than from fields. Total nitrogen released to the air in the form of ammonia (361,000 tons) is about twice the total quantity of manure nitrogen applied to crops and almost three times the quantity that is not absorbed by the crops. The high level of nitrogen released as ammonia implies that there is a significant potential for increasing manure nitrogen available for crops. We assume that both excess nitrogen and ammonia emissions in the baseline exceed environmental standards; i.e., further increases in either one would result in unacceptable degradations in environmental quality.

Wide geographical differences in application rates reflect the relative abundance of cropland on which manure is applied. Lagoon operations, located primarily in the Mid-Atlantic, apply manure nitrogen at 9.2 times the agronomic rate, on average, compared with 5.5 times the agronomic rate for pit operations, located primarily in the Corn Belt regions. Livestock operations in the Corn Belt tend to be more integrated with crop production than elsewhere, so they generally have more cropland available on the farm for spreading manure (McBride and Key, 2003). For pit operations, the amount of excess nitrogen applied to land that is not absorbed by crops is about the same amount of nitrogen released to the air as ammonia. Lagoon operations, in contrast, release far more nitrogen into the air, primarily from the lagoon itself.

Nitrogen Application Standard

The CAFO rules require farmers to follow a nutrient management plan that eliminates excess applications of nitrogen. In our first scenario, we assume each hog operator must meet a nitrogen application standard. Farmers adjust their operations to meet this standard at least cost. CAFOs increase the share of their own land on which they apply manure, decrease the share of the land cultivated using chemical fertilizer, and increase shipments of manure off-farm to conform to this standard. As a result, total profits from the hog enterprise and the whole farm (accounting for crop production) decline about 6.9 percent and 5.8 percent (table 3-1, column 2).

Economic impacts are not distributed equally between the major manure handling technologies. Hog profits for farms using slurry systems decline 9.9 percent, versus 4.9 percent for farms using lagoon systems (table 3.2, column 2). Pit operations suffer larger losses because slurry manure contains more nitrogen than lagoon liquid. Even though pit farms tend to have more land available for spreading manure, the high nitrogen content of the slurry means they must still transport large amounts of manure off the farm, and incur high hauling costs.

The nitrogen application standard effectively eliminates excess nitrogen applied to the soil. The nutrient application standard also induces a 3.4percent increase in the quantity of ammonia nitrogen emitted from fields (table 3-1, column 2), mainly because more land is receiving manure and because farmers that had been injecting slurry switch to surface application to use more manure and minimize off-farm transportation costs. However, the net effect of the policy on ammonia nitrogen emissions is very small, due mainly to the small decline in hog production. Hog production on each farm declines because of the increase in production costs relative to market prices for hogs (which are assumed constant). There is no real tradeoff between air quality and nitrogen available for crops because farmers were generally not taking steps to preserve the nutrient content of manure by preventing atmospheric losses in the first place. This suggests that large hog producers treat manure as a waste to be disposed of rather than a valuable source of nutrients.

Table 3-1. Production, profits, emissions, and technology adoption under nitrogen application standard (NAS), ammonia nitrogen standard (ANS), and both

Item	Base	NAS		ANS		NAS+ANS	
			% chg.		*% chg.*		*% chg.*
Hogs (*mil. cwt.*)	119.10	117.96	-0.96	118.26	-0.70	115.61	-2.93
Total profits (*mil. $*)	3,700	3,487	-5.77	3,426	-7.40	3,187	-13.87
Hog enterprise profits (*mil. $*)	3,047	2,837	-6.89	2,805	-7.93	2,568	-15.72
Ammonia N storage (*1,000 tons*)	327.5	325.3	-0.68	203.3	-37.91	198.8	-39.29
Ammonia N field (*1,000 tons*)	33.8	34.9	3.38	53.1	57.16	52.1	54.15
Ammonia N total (*1,000 tons*)	361.3	360.2	-0.30	256.4	-29.02	250.9	-30.55
Excess N soil (*1,000 tons*)	137.7	0.0	-100.00	246.4	78.95	0.0	-100.00
Application rate (*factor of agronomic rate*)	7.3	1.0	-86.38	17.6	140.37	1.0	-86.38
Manure transport costs (*mil. $*)	0.0	205.6	-	0.0	0.00	231.9	-
Manure N on-farm (*1,000 tons*)	183.6	51.8	-71.81	284.6	55.02	42.3	-76.96
Manure N off-farm (*1,000 tons*)	0.0	127.7	-	0.0	0.00	235.7	-
Cover lagoon (*% farms, all farms*)	0.00	0.00	0.00	36.42	-	36.42	-
Inject manure (*% land, all farms*)	25.56	22.55	-11.78	37.66	47.33	37.46	46.54

Ammonia Nitrogen Limit

We also consider ammonia nitrogen limits based on the minimum levels obtainable employing currently available abatement technologies (lagoon covers and manure injection). For this policy simulation, manure nutrient application standards for protecting water quality are assumed not to exist. For pit operations, ammonia nitrogen emissions are constrained to 10 percent above the minimum obtainable if all manure is injected. For lagoon operations, ammonia emissions are constrained to 20 percent above what is obtainable if lagoons are covered. These limits were chosen so that costs to producers of meeting the emission standards are in the same range as under CAFO application standards and so that pit and lagoon operations face similar regulatory costs under the joint policy (next scenario). The constraint on ammonia emissions is in the form of a percentage reduction in net N emissions per pig.[2]

The ammonia nitrogen standard induces pit operations to switch from surface application to injection on some land, and induces some lagoon operations to cover their lagoons. The ammonia standard results in a 38percent decline in ammonia emissions from manure storage facilities (the largest source of emissions) and a 57-percent increase in emission from fields, for a net decline in ammonia emissions of 29 percent (table 3-1, column 5). The increase in emissions from fields results because more lagoons are covered, which raises the nutrient content of the lagoon liquid applied to fields, resulting in greater nitrogen volatilization. The ammonia standard resulted in a 79-percent increase in excess nitrogen applied to soil—revealing an important tradeoff between water and air quality. For pit operations, the standard does not affect the profitability of the hog enterprise, so there is no hog production response (table 3-2). On the other hand, profits for operations with lagoons decline over 13 percent, resulting in a decline in production of about 1.2 percent, with no marketwide price effects accounted for.

Additional Simulations

To explore the tradeoffs between water and air emissions in more detail, we perform two more simulations. First, we examine how the levels of excess soil nitrogen and ammonia nitrogen vary for different nitrogen application standards applied to CAFOs (figure 3-1). The application standard is incrementally tightened from 50 percent greater than the agronomic rate to full implementation (where manure must be applied at the agronomic rate for all crops). Reducing the allowable nitrogen application rate (moving from right to left along the x-axis) results in a large decrease in the excess nitrogen applied to the soil, but almost no change in the amount of ammonia released. Farms were not conserving manure nutrients prior to the CAFO regulations, so the nitrogen application standards had little impact on air emissions. With no ammonia limits, tightening the nitrogen application standards to improve water quality produces only a minimal tradeoff in terms of lower air quality.

Table 3-2. Production, profits, emissions, and technology adoption under nitrogen application standard (NAS), ammonia nitrogen standard (ANS), and both, by storage technology

Item	Base	NAS		ANS		NAS+ANS	
			% chg.		*% chg.*		*% chg.*
Lagoon operations							
Hogs (*mil. cwt.*)	70.76	70.52	-0.34	69.92	-1.18	68.50	-3.20
Total profits (*mil. $*)	2,019	1,929	-4.47	1,778	-11.95	1,686	-16.50
Hog enterprise profits (*mil. $*)	1,827	1,738	-4.88	1,586	-13.22	1,494	-18.26
Ammonia N storage (*1,000 tons*)	255.0	254.1	-0.34	130.8	-48.69	128.2	-49.74
Ammonia N field (*1,000 tons*)	14.8	14.7	-0.34	40.7	175.05	39.8	169.43
Ammonia N total (*1,000 tons*)	269.8	268.8	-0.34	171.5	-36.43	168.0	-37.73
Excess N soil (*1,000 tons*)	42.4	0.0	-100.00	136.3	221.64	0.0	-100.00
Application rate (*factor of agronomic rate*)	9.4	1.0	-89.31	25.4	171.11	1.0	-89.31
Cover lagoon (*% farms, lagoon farms*)	0.00	0.00	0.00	76.70	-	76.70	-
Pit operations							
Hogs (*mil. cwt.*)	48.34	47.44	-1.86	48.34	0.00	47.11	-2.54
Total profits (*mil. $*)	1,681	1,558	-7.33	1,648	-1.94	1,501	-10.71
Hog enterprise profits (*mil. $*)	1,220	1,099	-9.91	1,220	0.00	1,074	-11.91
Ammonia N storage (*1,000 tons*)	72.5	71.2	-1.86	72.5	0.00	70.7	-2.54
Ammonia N field (*1,000 tons*)	19.0	20.2	6.27	12.4	-34.54	12.3	-35.52
Ammonia N total (*1,000 tons*)	91.5	91.4	-0.17	85.0	-7.17	82.9	-9.39
Excess N soil (*1,000 tons*)	95.3	0.0	-100.00	110.2	15.53	0.0	-100.00
Application rate (*factor of agronomic rate*)	5.5	1.0	-81.88	10.7	93.27	1.0	-81.88
Inject manure (*% land, pit storage farms*)	48.67	42.93	-11.79	71.70	47.31	71.32	46.54

A much different outcome occurs if hypothetical restrictions are placed only on ammonia emissions and CAFOs do not have to meet nutrient application standards (figure 3-2). A significant tradeoff between water and air quality would occur with increasing restrictions on ammonia emissions and no restrictions on nutrient application rates. Moving toward full implementation of emissions-reducing technologies (lagoon covers and manure injection) causes a large increase in excess soil nitrogen. Since there is no incentive to apply at agronomic rates, and spreading on more land would increase costs, manure nutrients are overapplied.

With no ammonia limits, tightening the nitrogen application standards to improve water quality produces only a minimal tradeoff in terms of lower air quality.

Coordinated (Air and Water) Environmental Policies

Relative to either of the single-medium policies, the joint policy is quite costly in terms of profits (column 7, tables 3-1 and 3-2). Hog operation and total farm profits decline by 15.7 percent and 13.9 percent relative to the base year. Production decreases about 3 percent. However, this policy reduces ammonia nitrogen by about 30 percent relative to the levels under the CAFO regulations alone, and eliminates excess nutrient applications. Lagoon operations suffer the largest decline in profits, with hog profits declining 18.3 percent, compared with 11.9 percent for pit operations.

An indication that the individual policies could be in conflict is that the least-cost mix of technologies for addressing either an ammonia policy or a water policy is not the same as the mix of technologies that best meets a joint policy goal. Meeting the land application goal results in a decrease in the amount of manure that is injected, contrary to what would be required to meet the ammonia emission reduction goal. Similarly, meeting the ammonia reduction goal would result in a large increase in nitrogen applications that are in excess of crop needs.

This result implies that applying one policy after the other would result in higher costs than applying both simultaneously. To meet the ammonia emission goal after first implementing nitrogen application standards, manure injection on pit operations would have to increase 66 percent, rather than the 46 percent if the goals had been met simultaneously. This additional 20 percent of land requiring injection is the direct cost of implementing policies piecemeal rather than jointly. This land had been injected before the implementation of any policy, but shifted to surface application in response to the nutrient application regulations. To shift back to injection imposes a cost that would have been avoided if the policies had been implemented simultaneously. For lagoon operators, this conflict does not occur. Meeting the nutrient application standards first would not require any subsequent changes in storage technology to meet the coordinated policy goals (lagoons remain uncovered).

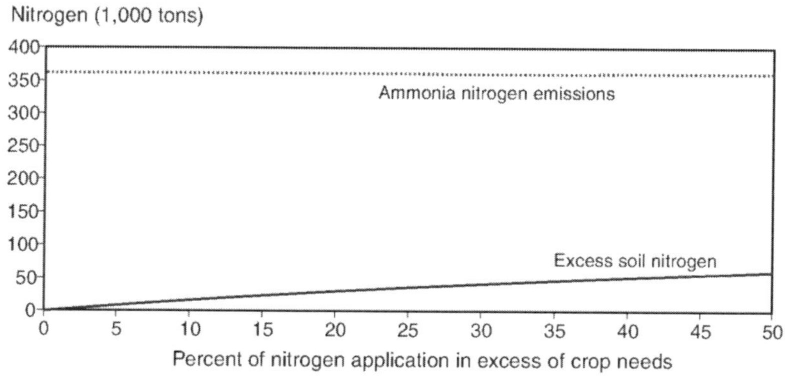

Note: Ammonia emitted to the atmosphere measured in terms of nitrogen.

Figure 3-1. Tradeoff between ammonia nitrogen emissions and excess soil nitrogen under varying soil nitrogen standards.

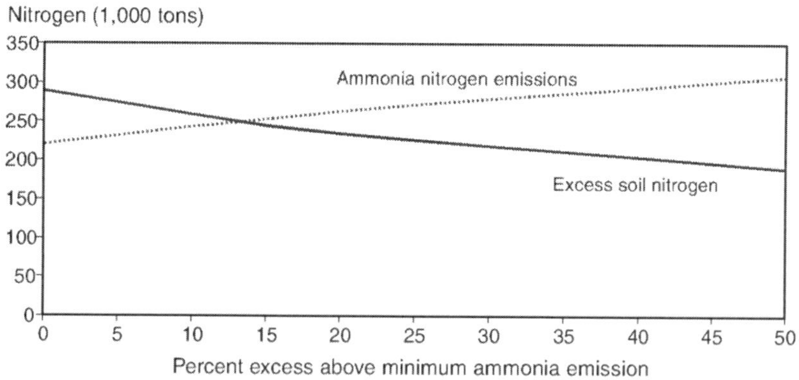

Note: Minimum ammonia emission based on all operations adopting emissions-reducing management practices.

Figure 3-2. Tradeoff between ammonia nitrogen emissions and excess soil nitrogen under varying ammonia nitrogen standards.

Farm-Level Decisions Have National Implications

This farm-level analysis highlights the economic and environmental tradeoffs that can occur with single-medium environmental policies as they are applied to hog farms. The two single-medium policies induce different responses. An ammonia emission standard alone would induce farmers to apply more excess nitrogen to the soil—a result likely to diminish water quality through increased nitrogen runoff and leaching. By itself, a nitrogen application standard to protect water quality does not have as dramatic an impact on ammonia emissions, but it does encourage an increase in surface application of manure, which increases potential ammonia losses from fields. Meeting both excess nitrogen and ammonia standards would be more costly than either single-medium policy.

Decisions made at the farm level are just the start of policy impacts. An operator who adopts waste management practices in response to regulatory requirements may also see an increase in production costs. Increased costs of production would, in turn, reduce the number of animals produced. When many hog farmers are affected by a policy change, new production levels may alter the market price of animal products and inputs (feedgrains, for example). Price changes springing from environmental regulations affect consumers and other sectors of the economy, and may cause animal producers to make further changes in their operations.

Section 4. National Effects of Coordinated Manure Management

Having examined the implications of addressing nitrogen concerns over water and air quality for one farm-level sector, we now take a larger view. The implications of coordinated (air and water) policies across regions and animal/crop sectors must account for interactions between crop and animal production and their subsequent economic and environmental impacts. Here, potential changes in commodity prices and shifts in production among regions are estimated assuming (1) adoption of land application standards for manure generated on concentrated animal feeding operations (CAFOs) and/or (2) reductions in nitrogen emissions to air from manure generated on all animal feeding operations (AFOs). Some consequences of national policy can only be viewed at this scale of analysis. Tradeoffs are not limited to the farm, but extend to regions and consumers. Market adjustments can produce contrary outcomes, even without the complication of conflicting single-medium policies.

At any level of analysis, adjustments to environmental policies entail increased costs to the producer. The magnitude of this increase depends on a number of factors, including the amount of manure transported for application, the availability of cropland for applying manure nutrients, the willingness of crop producers to substitute manure nutrients for commercial fertilizer, and regional heterogeneity in crop and animal production. Again, we consider current manure spreading regulations both independent of and in coordination with potential ammonia emission regulations across regions and sectors.[3] Specifically, we assess the environmental and economic implications of:

(1) Impacts of the 2003 Clean Water Act regulations for the spreading of animal manure on cropland for CAFOs (*Water*);
(2) Hypothetical reductions in atmospheric nitrogen emissions from animal feeding operations by 10-40 percent in the absence of manure nutrient application standards (*Air10, Air20, Air30,* and *Air40*); and
(3) CAFO water quality regulations plus AFO nitrogen emission regulations (10-40 percent reductions) in each region (*WaterAir10, WaterAir20, WaterAir30,* and *WaterAir40*).

We are looking for tradeoffs associated with implementing policies piecemeal rather than jointly, and indications that the different environmental goals (air quality vs. water quality) move the animal sector along different adjustment paths. Hence, we look at ammonia restrictions in the absence and presence of existing CAFO regulations for the protection of surface-water quality.

Simulating Coordinated Environmental Policies

We use the U.S. Regional Agricultural Sector Model (see Appendix B, web only) to assess secondary price and quantity effects between crop and animal production (USMP; House et al., 1999) at the national and regional levels (figure 4-1). We simulate restrictions on manure nutrient use on cropland and on nitrogen emissions from animal production.[4] The model estimates nitrogen emissions to the atmosphere, which allows us to constrain ammonia emissions directly. This model has also been used in previous analyses of the Clean Water Act (Ribaudo et al., 2003; Kaplan et al., 2004).

Table 4-1. Policy scenarios for simulation analysis

Scenario	Manure nutrient spreading standards for CAFOs	Reduction in nitrogen emissions for AFOs
Base	No	No
Water	Yes	No
WaterAir10	Yes	10%
WaterAir20	Yes	20%
WaterAir30	Yes	30%
WaterAir40	Yes	40%
Air10	No	10%
Air20	No	20%
Air30	No	30%
Air40	No	40%

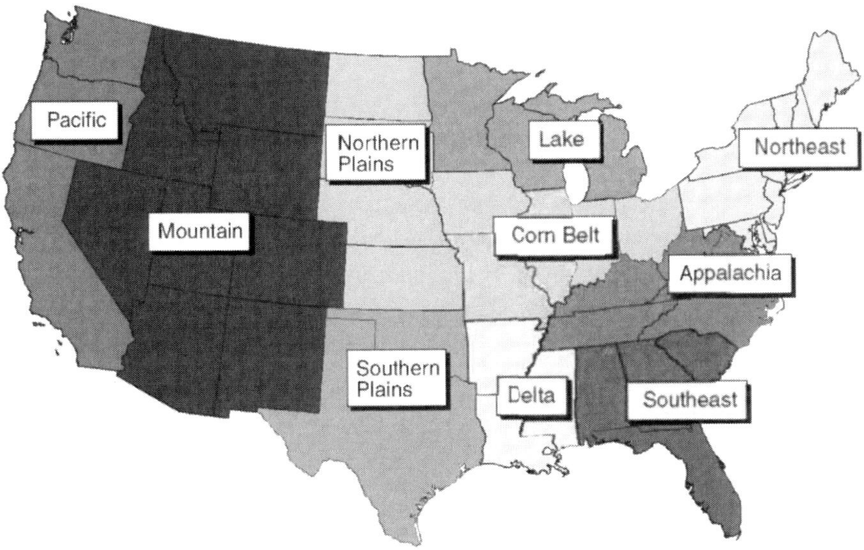

Source: Economic Research Service, USDA.

Figure 4-1. USDA farm production regions.

Various crop rotation, tillage, production and technology adjustments can be made to meet the nitrogen application or ammonia emission constraints. The composition of cropping or animal production could change to alter the amount of manure nutrients demanded or supplied. Storage, handling, or application technologies can reduce ammonia emissions or alter the nitrogen content of manure. Our model selects the optimal combination of technology, crop, and animal changes across the sectors and regions in order to minimize the net cost to society of meeting the different environmental policies. This includes changes in net returns for producers and changes in consumer surplus for purchasers of agricultural products. Storage, handling, and application technologies available in the model for meeting the CAFO nutrient standards and for reducing AFO emissions of nitrogen are consistent with those in the farm-level analysis.[5] We also consider treatment of poultry litter with aluminum sulfate (alum) to reduce nitrogen storage losses and to decrease the bioavailability[6] of phosphorus. Our baseline for comparison (*Base*) uses the USDA 2010 baseline projections for prices and production (USDA, WAOB, 2003) (table 4-1).

What Might We Expect?

CAFOs represent 4.5 percent U.S. feeding operations, but the quantity of manure generated by these facilities exceeds 200 million tons—more than 47 percent of the U.S. total (table 4-2). While the Corn Belt has the most AFOs and CAFOs and generates the most manure, the concentration of CAFO manure per cropland acre is greatest in the Appalachia, Southeast, and Pacific regions. Therefore, we would expect land application standards for CAFO manure nutrients to result in greater production adjustments in these relatively land-scarce regions. On the other hand, because all animal feeding operations are subject to ammonia emission policies in our analysis, regions with large numbers of animals, such as the Corn Belt, are likely to be more affected by such policies.

What We Found

Higher production costs from meeting environmental standards result in changes to production levels, both animal production (figure 4-2) and cropped acres (figure 4-3). Animal production would fall under all scenarios, but to different degrees across sectors.[7] Dairy production would remain relatively unchanged (reductions of less than 1 percent), but reductions in beef production would range from 0.1 percent (*Air10*) to 6.1 percent (*Air40*). Hog production would also exhibit a wide range of production changes.

Beef and hogs exhibit greater reductions because the cost of meeting some environmental constraints are higher. Many hog CAFOs are in regions with limited land for spreading manure (such as Appalachia), and would require expensive emission abatement technologies (covering lagoons). Much of feedlot beef production is in regions where crop demand for nutrients would be relatively low (such as the Southern Plains); under a nutrient application standard, adjustment costs are high. On the other hand, beef producers could meet the ammonia emission constraint-only scenarios (*Air10* through *Air40*) at relatively low cost. The combined scenarios (*WaterAir10* through *WaterAir40*) would be the most costly, primarily because of the high transportation costs borne by CAFOs to meet the land application requirements when atmospheric emissions are constrained, increasing the nitrogen content of manure.

Dairy CAFOs generally have more land available for spreading, which keeps hauling costs down. The ammonia abatement measures for poultry are generally less costly than for other sectors. In addition, poultry litter has a higher nutrient value-to-weight ratio, so it would be less costly to haul.

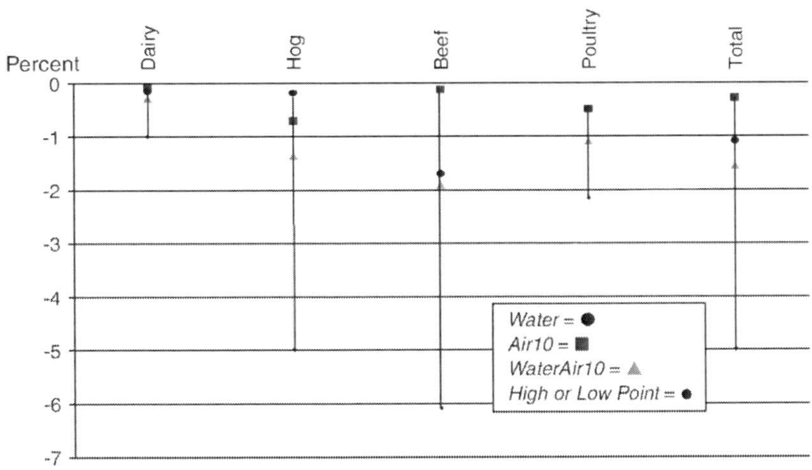

Figure 4-2. Changes in animal production by policy.

Table 4-2. Baseline for policy simulations

Region	AFOs						CAFOs		
	Operations	Total manure	Nitrogen runoff	Nitrogen leached	Nitrogen emissions	Phosphorus runoff	Operations	Manure	
							Share of total	Share of total	
		Million tons	*Million pounds*				*— Percent —*		*Tons/ acre*
NE	31,350	39.10	32.60	0.17	189.61	5.46	1.59	16.45	0.45
LS	52,498	61.54	72.50	0.80	362.55	12.30	1.64	26.68	0.43
CB	71,252	83.75	87.60	1.02	517.85	25.98	3.18	39.49	0.34
NP	26,087	71.13	80.05	0.82	371.35	15.42	4.77	62.91	0.64
AP	22,776	78.32	120.49	2.01	571.36	34.81	7.46	65.95	2.88
SE	12,635	24.35	126.97	0.67	187.64	21.87	10.97	43.48	1.43
DS	12,252	19.66	33.33	0.34	137.95	9.86	7.48	39.44	0.44
SP	10,500	48.42	72.65	0.43	263.98	17.01	7.00	38.77	0.62
MT	7,780	33.52	80.45	0.09	215.23	14.69	8.43	70.22	0.89
PS	7,654	39.53	118.55	0.18	283.10	16.40	14.85	60.49	2.50
US	254,784	499.31	825.19	6.52	3,100.62	173.78	4.47	47.44	0.72

Northeast (NE) = CT, DE, MA, MD, ME, NH, NJ, NY, PN, RI, VT; Lake States (LS) = MI, MN, WI; Corn Belt (CB) = IA, IL, IN, MO, OH; Northern Plains (NP) = KS, ND, NE, SD; Appalachia (AP) = KY, NC, TN, VA, WV; Southeast (SE) = AL, FL, GA, SC; Delta States (DS) = AR, LA, MS; Southern Plains (SP) = OK, TX; Mountain (MT) = AZ, CO, ID, MT, NM, NV, UT, WY; Pacific States (PS) = CA, OR, WA).

Table 4-3. Net manure storage, handling, treatment, hauling, and application costs

Region[1]	Water	Water Air10	Water Air20	Water Air30	Water Air40	Air10	Air20	Air30	Air40
	$ million								
NE	9.28	40.40	46.62	53.12	59.52	2.51	5.58	20.53	36.22
LS	34.55	77.76	92.87	110.90	127.56	3.97	8.03	20.41	58.08
CB	52.33	138.30	161.11	193.86	212.70	8.77	21.43	73.79	142.73
NP	129.26	127.38	137.11	132.13	201.74	2.78	9.38	33.20	139.40
AP	6.08	34.73	57.79	95.40	161.42	8.20	18.80	37.54	71.58
SE	23.24	29.33	41.64	53.96	66.75	5.53	16.47	27.93	39.65
DS	13.70	25.71	33.28	41.74	55.50	5.12	13.82	23.02	34.73
SP	47.35	47.65	56.24	74.11	54.49	1.73	7.05	16.68	61.33
MT	96.99	97.14	97.10	99.32	92.66	1.36	3.14	5.44	5.57
PS	121.69	124.00	124.37	126.57	128.76	2.27	4.87	7.73	17.15
Total U.S. cost	534.46	742.39	848.12	981.11	1,161.10	42.23	108.56	266.27	606.42
U.S. cost per AU ($)	4.99	6.96	7.96	9.31	11.19	0.39	1.01	2.52	5.89

[1] Northeast (NE) = CT, DE, MA, MD, ME, NH, NJ, NY, PN, RI, VT; Lake States (LS) = MI, MN, WI; Corn Belt (CB) = IA, IL, IN, MO, OH; Northern Plains (NP) = KS, ND, NE, SD; Appalachia (AP) = KY, NC, TN, VA, WV; Southeast (SE) = AL, FL, GA, SC; Delta States (DS) = AR, LA, MS; Southern Plains (SP) = OK, TX; Mountain (MT) = AZ, CO, ID, MT, NM, NV, UT, WY; Pacific States (PS) = CA, OR, WA.

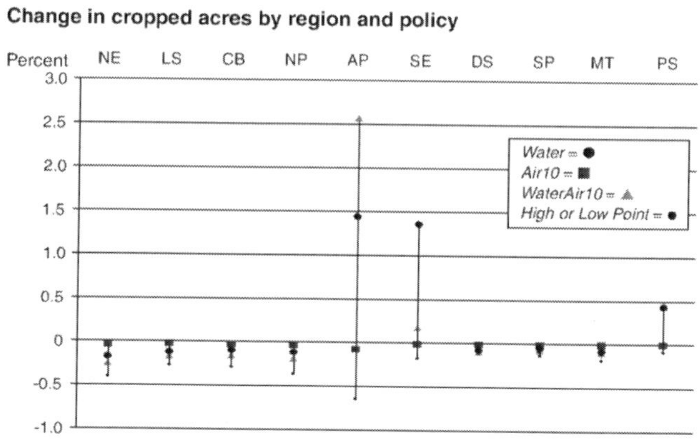

Northeast (NE) = CT, DE, MA, MD, ME, NH, NJ, NY, PN, RI, VT; Lake States (LS) = MI, MN, WI; Corn Belt (CB) = IA, IL, IN, MO, OH; Northern Plains (NP) = KS, ND, NE, SD; Appalachia (AP) = KY, NC, TN, VA, WV; Southeast (SE) = AL, FL, GA, SC; Delta States (DS) = AR, LA, MS; Southern Plains (SP) = OK, TX; Mountain (MT) = AZ, CO, ID, MT, NM, NV, UT, WY; Pacific States (PS) = CA, OR, WA).

Figure 4-3. Change in cropped acres by region and policy.

At greater reductions in air emissions, the combined scenario (e.g., *WaterAir40*) can have a smaller impact on costs in all animal sectors than the air-only scenario (e.g., *Air40*) because the increased nutrient value of manure would increase the amount AFOs receive for their manure from crop producers, mitigating land application costs (crop producers are assumed to pay for manure).

Under two scenarios (*Water* and *WaterAir10*), the incentive to increase cropped acres where the CAFO manure application standard is most binding (Appalachia, Southeast, and Pacific) boosts cropped acres by 0.5 to 2.5 percent (figure 4-3). For all other scenarios, cropped acres would fall as demand for feedgrain declines with decreasing animal production.

Technology Adjustments

Technology adoption would be influenced by the policy requirements and the relative costs of the management practices, determined by factors such as mix of animals, dominant production technologies, and cropland available for spreading manure. Consequently, total regional expenditures on practices for meeting environmental goals would adjust after simulated adjustments in production levels have taken place (table 4-3). Under the *Water* scenario, where only the CAFO regulations for protecting water quality are simulated,

CAFOs would develop and implement nutrient management plans that minimize the cost of spreading manure. Hauling manure to cropland would be the predominant cost. If restrictions are placed on ammonia emissions from animal feeding operations, producers would begin using alum, incorporating/injecting manure, and covering their lagoons. The costs of these alternative storage, handling, treatment, and application technologies would increase as required reductions in ammonia emissions increase. Expenditures would be highest when both air and water quality goals have to be met (*WaterAir10-40*) because more actions to manage manure must be taken.

The producers' cost per animal unit for each scenario reveals the same relationship between the air and water scenarios as in the farm-level analysis. The sum of the costs of the CAFO regulation scenario (*Water*) and the hypothetical ammonia-only regulation scenarios (*Air*) would be less than the cost of the joint policy scenarios (*WaterAir*) that achieve the same level of ammonia reduction. Again, this indicates that the policies are in conflict. The optimal economic adjustments to the different environmental policies would involve tradeoffs; i.e., strategies for addressing ammonia control may make nutrient standards for CAFOs more costly and vice versa.

Regional Shifts

Examining the pattern of geographic shifts in production following our policy simulations provides further evidence that addressing ammonia and water goals independently rather than jointly would impose additional costs on producers. These shifts result from many simultaneous economic forces, reflecting relative costs of meeting regulations, animal mix, and resource base. As expected, meeting two environmental goals rather than one would impose additional costs on the sector and would result in a larger reduction in production. The CAFO regulations alone could reduce production by about 1.2 million animal units (table 4-4). Simultaneously reducing ammonia losses by 20 percent could reduce production by an additional 650,000 animal units (a total loss of 1.85 million).

Most regions would follow this same pattern; production losses would be greatest when both environmental regulations are in place. However, in the Mountain, Appalachian, Northern Plains, and Pacific regions, production would be higher under the joint regulations than under the *Water* scenario. In these regions, the costs associated with closing operations and losing production that would have occurred under the *Water* scenario could be avoided if the water and ammonia regulations are implemented simultaneously rather than independently.

Environmental Implications

Environmental impacts result from changing manure management at the farm level, but also from changing the number of animals raised in each region. As AFOs adopt storage, handling, treatment, and application technologies to reduce ammonia emissions, the reductions in nitrogen runoff (from land application constraints) would gradually diminish. Runoff could even increase above baseline levels, supporting the findings of the farmlevel analysis (figure 4-4). With nutrient application standards in place on CAFOs (*Water*), nitrogen runoff would fall by about 12 percent. As restrictions on ammonia emissions are imposed, nitrogen runoff could increase. When emissions are reduced 40 percent, runoff would be about 7 percent lower than the baseline. Many of the environmental gains attributable to the CAFO regulations would disappear, primarily because non-CAFO operations are not subject to land application constraints. Manure spread on land from these operations would have higher nitrogen content due to technological changes adopted to reduce air emissions. Not restricting land application to agronomic rates would therefore increase nitrogen runoff. Again, policies addressing pollution to only one medium could increase emissions to a different medium. Without land application restrictions in place at all, reducing ammonia emissions would increase overall nitrogen runoff, even with fewer animals.[8]

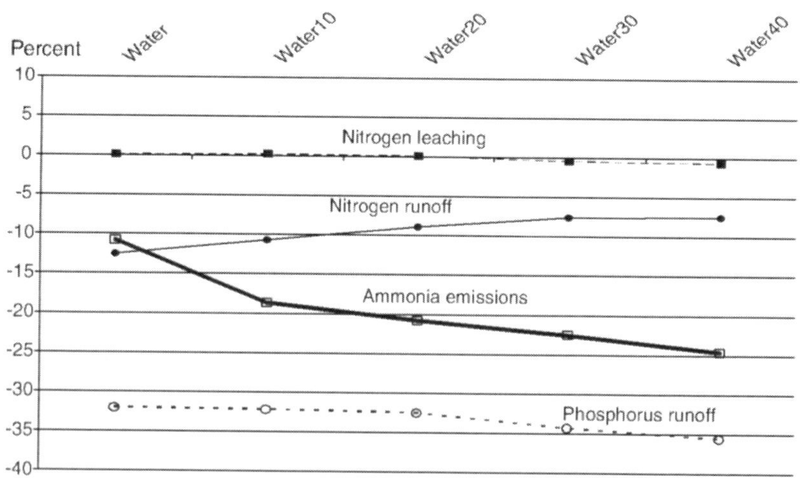

Figure 4-4. Change from baseline in ammonia and nutrient losses to the environment, U.S.

Phosphorus runoff is a leading cause of surface-water eutrophication (overenrichment of nutrients causing algal blooms). While phosphorus is not part of the nitrogen cycle, manure contains high levels of this nutrient and meeting a nitrogen standard may still result in an overapplication of phosphorus (Ribaudo et al., 2003). This can degrade water resources so that they are unfit for swimming, boating, or fishing. Manure phosphorus, along with nitrogen, is a focus of the Clean Water Act's CAFO regulations. Under most simulated policies, phosphorus discharges would fall substantially (figure 4-4). In the model, phosphorus applications are restricted under the CAFO regulations when phosphorus is the limiting nutrient (posing greatest environmental threat). Also, one of the practices for reducing ammonia emissions—adding alum to poultry litter—would further limit the loss of phosphorus to water.

Table 4-4. Changes in regional production

Region[1]	Base[2]	Water	Water Air10	Water Air20	Air10	Air20
	Million AU					
NE	4.176	-0.004	-0.123	-0.144	0.015	-0.068
LA	7.847	-0.099	-0.302	-0.359	0.045	0.110
CB	16.874	-0.375	-1.550	-1.725	-0.137	-0.251
NP	19.461	-0.848	-0.549	-0.648	-0.121	-0.643
AP	14.284	-0.323	-0.164	-0.225	-0.101	0.079
SE	3.871	0.005	0.019	-0.013	-0.043	-0.033
DL	3.082	-0.020	-0.120	-0.151	-0.053	-0.046
SP	21.224	0.400	0.729	0.880	0.042	-0.141
MN	10.365	0.450	0.651	0.755	0.041	0.254
PA	7.149	-0.358	-0.262	-0.220	0.006	0.082
US	108.333	-1.172	-1.671	-1.850	-0.306	-0.657

[1] Northeast (NE) = CT, DE, MA, MD, ME, NH, NJ, NY, PN, RI, VT; Lake States (LS) = MI, MN, WI; Corn Belt (CB) = IA, IL, IN, MO, OH; Northern Plains (NP) = KS, ND, NE, SD; Appalachia (AP) = KY, NC, TN, VA, WV; Southeast (SE) = AL, FL, GA, SC; Delta States (DS) = AR, LA, MS; Southern Plains (SP) = OK, TX; Mountain (MT) = AZ, CO, ID, MT, NM, NV, UT, WY; Pacific States (PS) = CA, OR, WA).

[2] Baseline values are taken from 2010 USDA baseline projections (USDA, 2003).

Economic Implications from Market Interactions

The farm-level analysis assumed constant prices, so the costs of meeting single-medium or joint policies would be borne fully by animal operations. However, animal production would be expected to fall under all scenarios (figure 4-2) as producers adjust to increased production costs brought about by more intensive manure management. Such changes would likely lead to higher commodity prices, transferring some of the burden of higher costs to consumers (table 4-5). Crop producers would also be affected by what happens in the animal sector. Corn and soybeans are important feedgrains for animal production, and fewer animals being produced would dampen corn and soybean prices. In addition, policies to protect water quality could increase corn and soybean acres in some regions (receiving manure), which would also reduce prices.

Economic tradeoffs from a joint rather than uncoordinated policy approach can only be inferred from our results. The livestock sector would seemingly benefit from a joint policy versus uncoordinated policy. Reductions in net returns for any one of the joint policies would be less than for the *Water* scenario. Costs to the livestock sector from implementing only the CAFO regulations could have been reduced if hypothetical ammonia reductions had been required at the same time.

Summary

A national manure management policy affecting a significant share of animal feeding operations would affect prices, producer net returns, and consumers. The environmental and economic impacts would vary greatly by region and animal type. The ability of the different sectors and regions to respond to the direct costs of water and air quality regulations depends on the size and structure of the agricultural operations, regional characteristics like available cropland, and responsiveness to price changes by the crop and animal sectors.

National results generally confirm farm-level results. Policies aimed at reducing ammonia emissions from animal manure would result in technological and production adjustments that could, in the aggregate, lead to increased discharge of nitrogen into surface and ground waters. This outcome is explored further for the Chesapeake Bay watershed.

Table 4-5. Economic impacts and nitrogen reductions

	Water	Water Air10	Water Air20	Water Air30	Water Air40
	Change (million units)				
Nitrogen reductions (lbs. runoff, leaching, and air emissions)	1,169	1,553	1,599	1,653	1,779
Net returns to crop production ($)	449	328	307	267	196
Net returns to live-stock production ($)	-897	-700	-724	-566	-268
Consumer surplus ($)	-402	-786	-876	-1,304	-2,053
Returns to agriculture and consumer surplus ($)	-850	-1,158	-1,293	-1,602	-2,125

SECTION 5. IMPACT OF SPATIAL FACTORS ON THE COSTS OF MANURE MANAGEMENT: A CASE STUDY OF THE CHESAPEAKE BAY WATERSHED

The costs associated with meeting USDA goals and U.S. EPA regulations for improved manure management depend not only on individual farm conditions and national markets, but on spatial considerations concerning the location of animal operations and cropland available for manure application. Where animal production is concentrated, producers may face competition for suitable land to apply manure, which can increase the cost of meeting application requirements by forcing manure to be hauled longer distances (Ribaudo et al., 2003). Implementing hypothetical ammonia emission controls on farms meeting nutrient standards could increase the competition for land by increasing the nutrient content of manure. Among U.S. areas where manure nutrient production exceeds the assimilative capacity of the land are several county clusters within the Chesapeake Bay watershed (Gollehon et al., 2001) (figure 5-1). We present this case study to demonstrate how the costs of meeting Clean Water Act requirements might be affected if ammonia emission regulations are also imposed.

The Chesapeake Bay is among the largest and most biologically rich estuaries in the world. The declining health of this ecosystem in recent decades has prompted Federal and State initiatives to reduce nutrient loading from tributaries that drain the watershed. Nutrient discharges to waters in the region have resulted in eutrophication and related ecological shifts that harm wildlife and aquatic resources (Preston and Brakebill, 1999). Manure from confined animal operations has been identified as a primary source of both nutrient runoff to water bodies and local air emissions (Follett and Hatfield, 2001). A joint effort by watershed States to reduce nitrogen loadings is addressing all sources of nitrogen. The potential cost is high. The plan to upgrade 66 of Maryland's major sewage treatment plants will cost between $750 million and $1 billion, and would provide only a third of the nitrogen reductions needed for Maryland to meet its commitment (Maryland Department of Natural Resources, 2004).

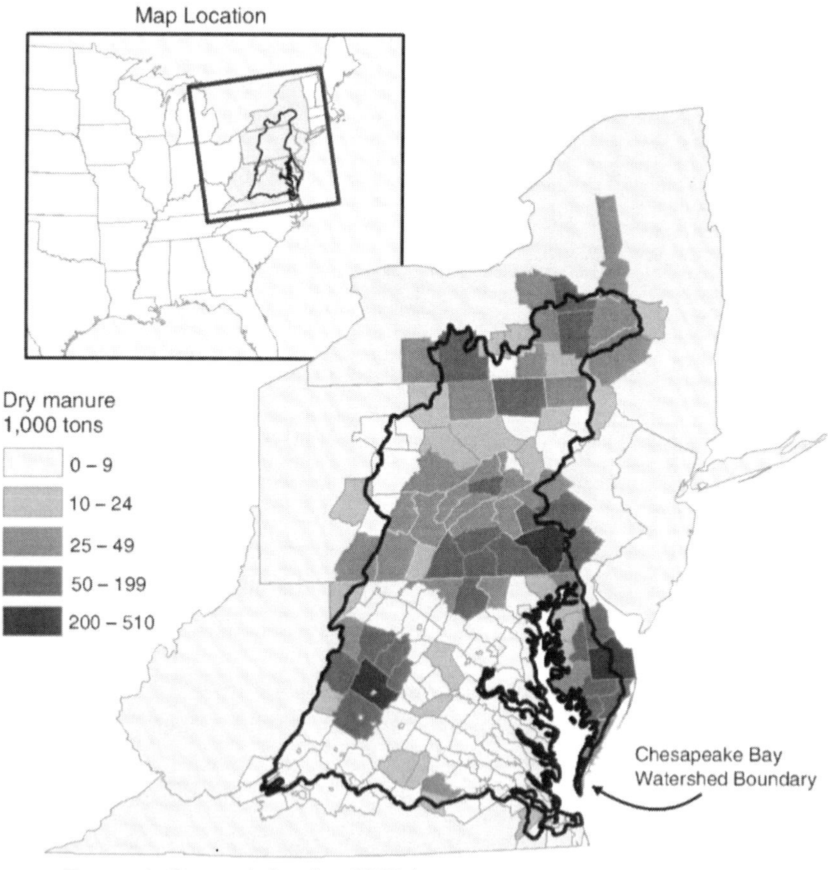

Source: Economic Research Service, USDA.

Figure 5-1. Location of manure production in Chesapeake Bay watershed, 1997.

The Chesapeake Bay watershed (CBW), spanning over 160 counties in 6 States, includes 66,600 farms with an estimated 8.5 million acres of land potentially available to receive manure. Approximately 15,900 farms in the CBW had confined animals in 1997, with an average daily inventory of about 1.6 billion pounds of feedlot beef, dairy, swine, and poultry (USDA, 1999). These animals produce roughly 93,000 tons of recoverable manure nitrogen, 44,000 tons of recoverable manure phosphorus, and 100,000 tons of ammonia-N annually. Even if confined animal operations fully utilized the crop and pasture land under their control for manure application, excess nutrients would remain. Applying manure at agronomic rates to meet water quality goals would require moving significant quantities of manure off animal producing farms.

The effect of alternative behavioral assumptions on the willingness of farmers to accept manure as a substitute for commercial fertilizer was examined in Ribaudo et al. (2003) for all confined animal operations in the CBW. Results from that study indicate that if farmland application is the primary disposal method, implementing nutrient management regulation poses significant challenges where animal production is concentrated. Only about half the manure produced CBW-wide can be used onfarm given current technologies and crop mixes. The feasibility of land application as a regional manure management strategy depends on the willingness of landowners to accept manure on farmland, the nutrient assimilative capacity of the regional cropland base, and the nutrient standard in effect. Ribaudo et al. estimated that more than 30 percent of CBW crop farms would need to accept manure in order to land-apply all the manure produced in the CBW at a rate based on the nitrogen needs of crops (under reasonable hauling distance assumptions).

The CBW case study uses a regional modeling framework designed to capture spatial considerations in manure production and land availability for manure spreading (see Appendix C, web only). The model and its results reflect a regional planning perspective in evaluating key cost determinants and alternative policy strategies at the watershed scale.

We assume that farms meeting nutrient application standards will apply manure at a rate based on a **nitrogen** standard for cropland and pastureland. Farms in locations with high phosphorus concentrations in the soil and runoff vulnerability may be required to base manure applications on a phosphorus standard, which generally decreases manure applied per acre (Ribaudo et al., 2003). While the effects of manure land application on phosphorus-limiting soils is an important concern in the Chesapeake Bay region, air emission controls interact primarily with manure-nitrogen concentrations. Thus, our focus is on changes in costs to meet a nitrogen standard.[9] We assume the willingness to accept manure by crop producers is 30 percent.

Determining the effect of emission control technologies on representative manure handling systems for the CBW required two steps. First, the quantities of total manure excreted were estimated from quantities of recoverable manure nitrogen available in the watershed (Aillery et al., 2005) and estimates of ammonia-N losses at the facility and field levels (appendix table D-1). This estimation process—from the field back to the animal—provides consistent estimates of manure nitrogen and ammonia emissions for a baseline situation. (Estimated values for selected systems commonly used in the CBW are provided in appendix table D-2.) In the second step, quantities of manure nitrogen available for plant use and changes in ammonia emissions were estimated from the animal to the field with the addition of emission control technologies by a manure-handling system. Nitrogen losses by system (see section 2) were converted to losses as a share of recoverable manure for direct inclusion in the model (appendix table D-3). An example of the estimation process to calculate nitrogen losses and crop availability under alternative emission control technologies is provided in appendix table D-4.

Land Application of Manure with Ammonia-N Reductions

We compare costs to the CBW animal sector of land-applying manure, and the water quality impacts, of four scenarios:

Case A—CAFOs meet nitrogen-based land application standards for water quality improvement, without consideration of ammonia-N emissions (current Clean Water Act policy);

Case B—CAFOs meet nitrogen-based land application standards for water quality improvement, with the addition of ammonia-N reducing technologies and practices;

Case C—All AFOs adopt ammonia-N emission controls for air quality improvements, while CAFOs continue to meet land application standards;

Case D—All AFOs adopt ammonia emission controls and meet land application standards.

For purposes of this analysis, methods of controlling ammonia emissions include the following:

- ***Incorporation/injection.*** Manure is incorporated or injected on 100 percent of acres receiving manure from poultry, dairy, and feedlot beef operations in the included farm set. We assume that lagoon liquid from dairies and feedlot beef operations is surface applied so that it is possible to inject the liquid with current technologies; swine lagoon waste is generally sprayed and is not typically incorporated. Under current conditions, incorporation is assumed to occur on 40 percent of CBW cropland for soil-nutrient retention and odor control, based on data from the ARMS hog and dairy surveys. This practice reduces ammonia emissions on acres currently treated. We assume the crop mix on land receiving manure does not change.
- ***Lagoon covers.*** Impervious lagoon covers are added to all dairy, swine, and feedlot beef operations using lagoon-based manure storage systems. The base model assumption is that no lagoons are covered.
- ***Alum.*** Alum is added to all poultry operations as an additive to the manure in the poultry house. The base model assumption is no alum use.

Case A—CAFO land application standards, with NO ammonia controls

The annual cost of meeting regulations for improved manure management to protect water quality is estimated to be $30 million[10] when only CAFOs meet land application standards (figure 5-2). CAFOs account for roughly 19 percent of the total modeled manure in the CBW. The distribution of CAFO farms varies significantly across CBW counties, with the share of AUs on CAFO operations ranging from 0 (for about half the counties) to as high as 80 percent in some counties. The manure from non-CAFO farms is assumed to be applied on the source farm without the benefit of a nutrient management plan. Ammonia-N emissions from manure produced on all AFOs in the watershed total an estimated 100,000 tons, including 78,000 tons from animal production and manure storage facilities and 22,000 tons from field applications (figure 5-3).

Case B—CAFO land application standards, with CAFO ammonia controls

The estimated cost to CAFOs for managing manure increases by $18 million relative to Case A (figure 5-2). This reflects both the cost of implementing ammonia-controlling practices ($9 million) and the increased cost of applying manure according to a nutrient management plan ($9 million).

Land application costs increase as a larger land area (more than doubled) is required to accommodate the nitrogen-enriched manure (figure 5-4).

The addition of ammonia emission controls on CAFOs would reduce emissions by about 12,000 tons, relative to Case A, representing 12 percent of total animal emissions basinwide (figure 5-3).

Case C—All AFO ammonia controls, with CAFO land application standards

Requiring all AFOs to control ammonia emissions would result in a CBW-wide reduction in emissions of 43 percent, 30,000 tons more than in Case B (figure 5-3). The additional cost to implement air emission controls through expanded use of alum, lagoon covers, and incorporation on all AFOs, relative to Case A, is an estimated $41 million (figure 5-2).

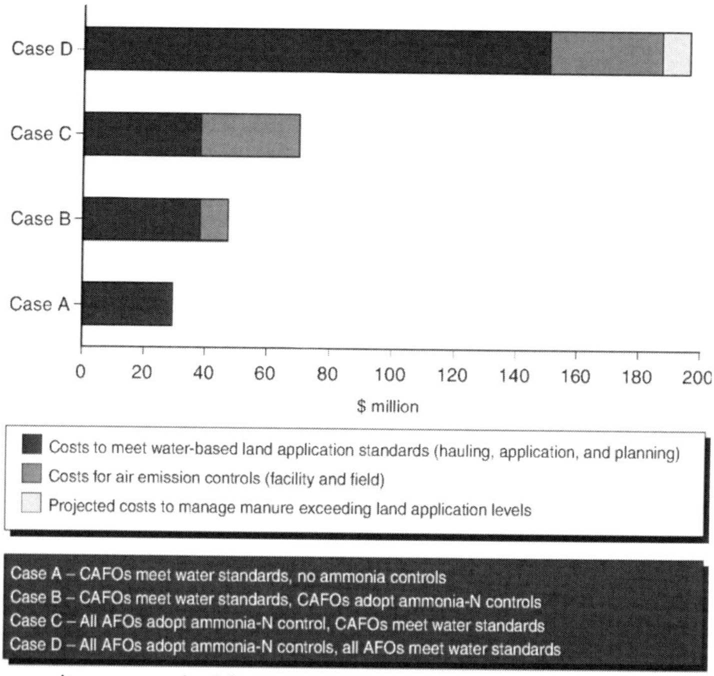

* Assumes a nitrogen standard for manure land application with 30 percent of farmland accepting manure.

Source: Economic Research Service, USDA.

Figure 5-2. Annual costs of meeting nutrient application standards with alternative emission controls on CAFOs and AFOs, Chesapeake Bay Watershed.*

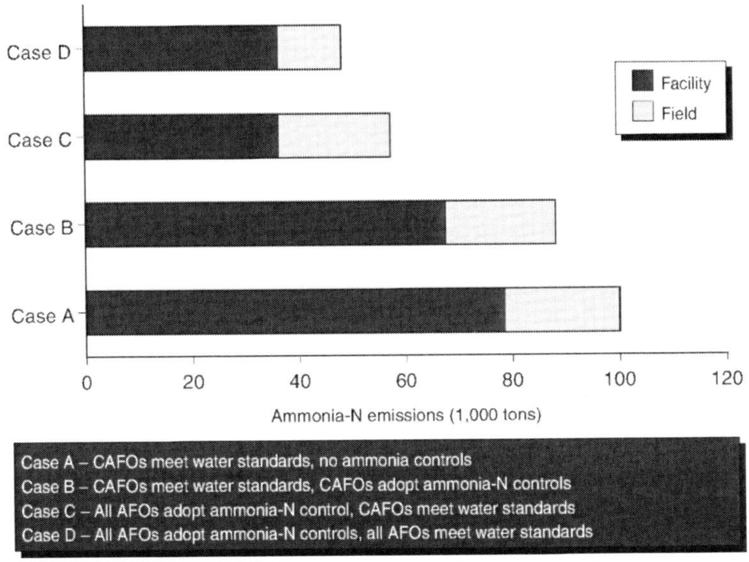

*Assumes a nitrogen standard for manure land application with 30 percent of farmland accepting manure.
Source: Economic Research Service, USDA.

Figure 5-3. Ammonia-N emissions with alternative emission controls for CAFOs and AFOs, Chesapeake Bay Watershed*

The threat of nitrogen runoff from CAFOs remains unchanged because the change in nitrogen content of manure is considered in the development of the nutrient management plans (figure 5-3). However, the same cannot be said for non-CAFOs, which are not required to follow a nutrient management plan. Non-CAFOs are assumed to spread their manure on land near the production facility without a nutrient management plan, at an estimated cost of $80 million (not shown in figure 5-2). If these operations do not adjust the amount of land receiving manure, the doubling of the nitrogen content of manure would increase the threat of runoff to the Bay.[11] Potential impacts on water quality would be an unintended consequence of the air quality policy if additional steps are not taken to address manure nutrient over-application.

Case D—All AFO ammonia controls, with AFO land application standards implemented simultaneously

A requirement that all AFOs follow a nutrient management plan in conjunction with meeting ammonia controls would likely limit the threat to water quality in the Bay while reducing ammonia emissions, but at an increased cost to producers. The decline in ammonia emissions is about 9,000 tons more than achieved under Case C (figure 5-3). The lower field emissions under Case D reflect the overall reduction in field losses achieved by applying all manure at agronomic rates, in contrast to Case C where most of the manure (81 percent) is assumed to be applied at rates substantially above crop needs. The total estimated cost for managing manure to meet ammonia control and land application requirements exceeds $186 million (figure 5-2). Most of this reflects the implementation of nutrient management plans on all AFOs and a 75-percent increase in the number of acres receiving manure in an environment of intense competition for land (figure 5-4). Land application costs on non-CAFO operations increase by an estimated $32 million.

An additional cost is for dealing with excess nitrogen for which no land is available within the CBW. Based on model simulations, the CBW has insufficient land to receive all manure when all AFOs are both controlling ammonia emissions and following a nutrient management plan, given an assumed farmer willingness to accept manure of 30 percent.

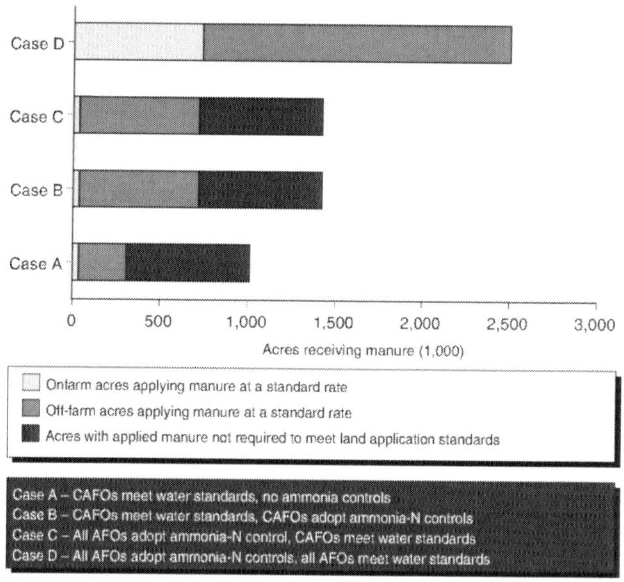

*Assumes a nitrogen standard for manure land application with 30 percent of farmland accepting manure.

Source: Economic Research Service, USDA.

Figure 5-4. Acres receiving manure from CAFOs and AFOs with alternative air emission controls, Chesapeake Bay Watershed*

Various disposal strategies exist for handling this basinwide surplus, like increasing the willingness of crop farmers to accept manure (assumed to be 30 percent in this case study), increasing manure diverted to industrial processes, adjusting the diet of animals to reduce manure nutrients, reducing the number of animals, or transporting manure beyond the 100-mile limit assumed in the model. The latter option may be the least expensive. States have recognized the need to move manure extended distances to comply with nutrient regulations, and some offer a transportation subsidy. Delaware, for example, provides transportation assistance of $18 per ton (Rohrer, 2004). For illustrative purposes, if we assume that all the surplus manure in the model could be transported to land outside the CBW for application and/or other disposal at $18 per ton, disposal of all AFO manure produced in the CBW would exceed $9 million annually (figure 5-2).

Reducing the region's animal units to a level where all manure could be land applied is a costly alternative. This approach would require as much as a 7percent reduction in the number of animals in the CBW. With an average regional return per animal unit of $1,339[12], the annual costs of reducing the number of animals, plus land application of the manure produced, would total about $150 million at a willingness to accept of 30 percent.

CONCLUSION

Our analysis brings to light a key challenge in achieving air and water quality goals: strides to meet one goal may impede the other, as nitrogen is either applied to farmland via manure or emitted to the atmosphere. Animal producers in the Chesapeake Bay watershed required to meet regulations and guidelines for water quality protection face significant costs for managing manure. New air emission controls, if implemented, would increase costs to the animal sector. The higher nitrogen content of applied manure would pose a challenge to those producers with limited land, requiring longer hauling distances to access adequate land. For other farmers needing nutrients, the nutrient-rich manure can be a resource.

Widely adopted ammonia emission controls could encourage overapplication of manure-nitrogen on non-CAFOs, to the detriment of water quality. Extending land application standards to non-CAFOs would substantially reduce nitrogen available for runoff, but substantially increase the total cost of air and water pollution abatement. The actual effect on water quality in the Bay will depend on nutrient loadings from applied (and overapplied) manure, and the deposition from airborne nitrogen.

Under current Federal regulations for land application of manure, the cost of air emission control reflects both the cost of control practices as well as the increased costs of meeting land application standards due to the higher nitrogen content of manure. Assessments that address practice implementation costs alone may substantially underestimate the full impact of air emissions control on the animal sector.

Cost impacts would be greatest where animal production is concentrated and manure quantities approach or exceed the assimilative capacity of the existing land base, increasing competition for land needed for manure spreading. Under these conditions, reliance on land application alone as a regional manure management solution may not be feasible. Other measures—such as increasing landowner willingness to accept manure, developing industrial applications for manure, subsidizing the long-range transport of manure out of the watershed, or even reducing animal stocks—may play a role in dealing with a regional surplus of manure nutrients.

SECTION 6. SUMMARY AND IMPLICATIONS FOR POLICY AND RESEARCH

Addressing the pollution problems generated by production activities can challenge policymakers concerned with economic efficiency when more than one environmental medium is affected by a single pollution source. This is true of many environmental issues. Coordinating policy so that all potential pollution issues are addressed simultaneously has been shown in the literature to be more efficient than dealing with each issue in an uncoordinated fashion, particularly when correcting one environmental problem worsens another (policies conflict). This chapter illustrates the tradeoffs in environmental quality by focusing on livestock and poultry production. Nitrogen in manure from animals on feeding operations can take a number of forms; reducing one form of nitrogen to protect one environmental medium can increase the amount of another form moving to a different medium.

Current environmental policies often fail to account for these interactions between media, as is the case with animal waste policies. Revised Clean Water Act regulations focus on managing land application of manure to reduce pollution of surface water. Restricting manure nutrient applications to agronomic rates can reduce manure's threat to water quality, but imposes costs on producers, primarily increased manure hauling costs. These costs can be reduced with manure handling and application strategies that promote the creation of ammonia and its loss to the atmosphere.

Ammonia emissions are a source of haze in the atmosphere, and a potential threat to human health. Recent lawsuits, court decisions, and consent agreements have induced some States to start regulating emissions under the Clean Air Act; the Comprehensive Environmental Response, Compensation and Liability Act; and State laws.

Tradeoffs between Air and Water Quality Are Prevalent in Manure Nitrogen Management

Tradeoffs between air and water quality exist at the farm, regional, and national level. However, the magnitude varies, depending on baseline assumptions and scale of analysis. The farm-level analysis found that CAFO regulations for reducing nitrogen runoff had only a small impact on ammonia emissions in the hog sector. Hog operations had generally adopted practices that released ammonia to the atmosphere prior to the CAFO regulations (open lagoons and surface manure application), so there was little opportunity to further increase these emissions. Model results suggest that some operations that had been incorporating slurry (presumably for odor control) would start surface-applying manure in order to reduce the amount of land needed to receive manure, but this effect was relatively small. At the national level, implementing the CAFO water quality regulations could actually reduce ammonia emissions. Estimated reductions in animal numbers outweighed any small increase in per-unit ammonia emissions.

On the other hand, when we assumed hypothetical restrictions on ammonia emissions in the absence of the CAFO regulations, excess nitrogen applications (the cause of nitrogen losses to water) increased dramatically at the farm level, and to a lesser degree at the regional and national levels. Animal operations reduced ammonia emissions by covering lagoons and incorporating manure. Excess manure nitrogen applications that existed in the baseline increased as nitrogen in manure was conserved by preventing losses to the atmosphere and land applied to the same acres.

Uncoordinated Policies Impose Extra Costs on Farmers

CAFO regulations and the hypothetical ammonia reduction regulations provide much different incentives to farmers, and so encourage different management practices. Furthermore, neither set of management practices is the most economical for addressing a joint policy where both water quality and ammonia emission goals are set. Farms that adopt a set of practices to meet the CAFO water quality requirements might need to adopt a different set to meet both water and air requirements. The cost of changing practices could be avoided under a coordinated policy. A producer may even be reluctant to comply with new regulations for fear that the rules may change in the future. (Our models do not account for uncertainty or for the economic implications of "sunk" costs for adopting a set of waste-handling technologies and then having to adopt a new set.)

These differences in production costs have broader implications for the agricultural sector. Overall impacts for animal producers, crop producers, and consumers change considerably when hypothetical ammonia reduction goals are added to the CAFO regulations. Some of the regional shifts in production predicted for the CAFO regulations do not occur under a coordinated policy. This implies that some of the adjustment costs from a sequential (uncoordinated) implementation of regulations would have been avoided if they had been introduced together.

Unintended Consequences Can Lessen Environmental Gains

Should ammonia emission standards induce farmers to adopt manure management practices that reduce nitrogen emissions, the manure applied to land will have a higher nitrogen content. Depending on how the air quality regulations are applied, this can have two impacts on CAFOs and water quality. First, those farms identified as CAFOs may need to increase the amount of land they are spreading on to meet nutrient application standards if they are also required to reduce ammonia emissions. This can be particularly costly in a region where animal concentrations are high and cropland available for spreading manure is relatively scarce. In our analysis of the costs of spreading manure in the Chesapeake Bay watershed, nitrogen content of manure substantially increased when ammonia restrictions were introduced, increasing the costs of meeting nitrogen application standards. The higher cost of meeting water quality regulations might not be accounted for in an assessment of the cost of air quality regulations.

Second, a failure to coordinate water and air policies could lead to a loss of water quality benefits. If Clean Air Act or CERCLA requirements result in States requiring ammonia reductions on smaller farms as well as CAFOs, the water quality benefits of the CAFO regulations could be diluted by excess nutrient applications on the smaller farms. This was the case in both our regional and national analyses. Without regulations on spreading manure at agronomic rates, farms reducing ammonia emissions would be more likely to overapply manure, thus increasing nitrogen discharged to surrounding waters. It would be difficult to achieve ammonia emission reductions and still maintain water quality gains of the CAFO regulations if water quality regulations were not extended to smaller operations. Doing so would increase the costs to producers and consumers, but provide greater environmental improvements.

In our analyses, we have not assumed a relative value of air quality changes versus water quality changes. Monetary values associated with improved health and visibility from reduced ammonia emissions, and improved recreation and drinking water benefits from reduced nitrate runoff, have not been estimated at the national scale. These values would help policymakers respond appropriately to ammonia emissions and nitrogen runoff from animal feeding operations. For example, if water quality improvements are valued much less than air quality improvements, the "unintended consequence" of increased nitrogen runoff from an uncoordinated ammonia control policy may be of little concern.

Other Tradeoffs May Be Important

While we focus on the ammonia-nitrate tradeoffs, other interactions have a bearing on current environmental concerns. For example, animal operations are a primary source of the greenhouse gases methane and nitrous oxide. The former is not part of the nitrogen cycle. Its sources are the animal itself (enteric processes) and anaerobic storage. Nitrous oxide, which is part of the nitrogen cycle, is emitted primarily from fields where nitrogen is applied and from dry-waste-handling systems that have aerobic conditions. Commercial fertilizer is the primary source of agricultural nitrous oxide.

Policies that influence the number of animals, manure handling and storage systems, and the amount of nitrogen applied to land also influence greenhouse gas emissions. For example, if a nitrogen runoff-ammonia scenario reduces the number of beef cattle by 2 percent, methane emissions from beef cattle also decline 2 percent. Considering potential conflicts and synergies between policies aimed at visibility, health, water quality, and global climate change would be complex and costly, but could avoid unnecessary costs to the sector and to society as a whole.

Reducing Nitrogen at the Source Can Address Multiple Problems

Not creating pollution in the first place avoids the problems posed by conflicting policies. In the case of nitrogen and AFOs, increasing the efficiency of nutrient conversion to animal products can reduce nitrogen in waste. This would reduce the threats to air and water quality, and make addressing either or both by managing manure less costly. Major advances are being made in feed efficiency through animal genetics, herd management, phase feeding, and feed formulation.

Another onfarm option involves technology for separating manure into liquid and solid wastes. Each component has a different nutrient content and may be handled differently so that the overall cost for meeting water quality and air quality goals may be reduced.

Another option is to remove manure as quickly as possible from the animal facility and to use it as an input elsewhere. Manure is currently being used to produce commercial fertilizer and energy, and research is underway to identify other potential uses. Atmospheric emissions may be more easily controlled in an industrial setting where contact with air and water can be minimized, and emissions from a ventilation system can be filtered. Current cost and demand conditions have not yet spurred wide-scale development of such industrial options. However, as environmental concerns increase and local, State, and Federal governments deal with manure issues, the costs of manure management are also likely to rise, making industrial options for manure more viable.

This chapter takes a stylized approach to mass balance, focusing on one set of compounds (nitrogen) and two environmental media (surface water and air quality). A more complete analysis would consider atmospheric deposition of nitrogen (bringing in nitrogen emissions from other sources), greenhouse gases, and groundwater contamination. A full accounting of all the controls necessary to meet additional environmental issues would likely increase the cost to producers, but the magnitude would depend on the interactions between the different pollutant flows, and the degree to which manure handling technologies can address multiple problems. However, the essential link between production and environmental quality, and the tradeoffs between different policy approaches, would likely be similar to those suggested by these results.

APPENDIX A. MODELING THE FARM LEVEL

Linear Program to Calculate Dual Values

Positive mathematical programming (PMP) is used to calibrate the farm-level model to base year data without having to add constraints that cannot be justified by economic theory. PMP takes advantage of the fact that it is easier to collect information about outputs and inputs at the farm level than information about costs. The observed output and input levels result from a complicated decision process based in part on a cost function that is known to the farmer but difficult or impossible to observe directly. Some costs—perhaps associated with the environment, risk, or technology—may be hidden to the researcher even when a detailed survey instrument is available. PMP incorporates information about unobservable costs by using a quadratic cost function that approximates the true underlying cost function.

There are three steps to the PMP calibration (Howitt, 1995). In the first step, a constrained linear programming model is used to derive dual values associated with the "calibration constraints." In the second step, the dual values are used to parameterize a calibrated quadratic objective function. In the third step, the calibrated model is used for economic analysis, by imposing environmental policy constraints.

In the first step, the linear objective is to maximize total net revenues:

$$\max_{X1_{ir}} \sum_{r} \sum_{i} X1_{ir}(P_{ir} - C_{ir}), \tag{1}$$

where $X1_{ir}$ is the level of each output i in region r. The cost of producing each output is $C_{ir} = \sum_{j} A_{ijr}W_{jr}$, where A_{ijr} is the amount of input j required to produce a unit of output and W_{jr} is the input price. The optimization is subject to $j \times r$ resource constraints:

$$\sum_{i} A_{ijr} X_{ir} \le \sum_{i} A_{ijr} X0_{ir}, \ \forall \ j,r \tag{2}$$

where $X\,0_{ir}$ is the initial observed activity level, so that $\sum_{j} A_{ijr}X0_{ir}$ is the initial level of input j.

Inputs include land, capital, feeder pigs, feed corn, feed soy, and chemical nitrogen fertilizer. Outputs include hogs, corn, soybeans, and "other crops" (defined as the value of all other crops produced). All three crops can be produced under three fertilization regimes: (1) chemical fertilizer, (2) manure fertilizer applied to the surface, or (3) manure fertilizer injected into the soil. We use the extension of PMP developed by Röhm and Dabbert (2003) to allow for a greater policy response between crop fertilization regimes than between crops. To do so, we define three "variant activities" (chemical fertilizer, manure-spread, and manure-injected) for each crop and impose calibration constraints that distinguish between variant activities and the total activity for each crop. In practice, this approach results in greater substitution between, for example, corn fertilized by spreading manure and corn fertilized by injecting manure, than between corn and "other crop" production.

The calibration constraints for each activity are:

$$X1_{ir} \le X0_{ir}(1+\varepsilon_1), \ \forall \ i,r \quad dual: \hat{\lambda}_{i,r} \tag{3}$$

where ε_1 is a small perturbation (see Howitt, 1995). Following Röhm and Dabbert, we include three additional calibration constraints corresponding to each set of variant activities. For corn activities, the additional calibration constraint is:

$$\sum_{i \in cv} x1_{ir} \le \sum_{i \in cv} x0_{ir}(1+\varepsilon_2), \ \forall \ i,r \quad dual: \hat{\lambda}_{corn,r} \tag{4}$$

where cv is the set of corn variant activities: cv = {corn chemical fertilizer, corn spread manure, corn injected manure}. There are two additional constraints analogous to equation 4 corresponding to soybean variant activities, sv, and other crops variant activities, ov.

From the 1998 ARMS survey and other sources, we observe prices P_{ir}, W_{ir}, the output levels $X0_{ir}$, and most of the input-output coefficients A_{ijr} (see Appendix tables A1-A4 for details). It would be desirable to include manure nitrogen as an input. However, we do not observe manure application rates, only the amount of land on which manure is applied.

Appendix table A-1. Initial production, $X0_{ir}$

Outputs	Units	Value	Source
Corn fertilizer	100 bushels	*	USDA ARMS Survey 1998
Corn manure surface	100 bushels	*	USDA ARMS Survey 1998
Corn manure inject	100 bushels	*	USDA ARMS Survey 1998
Soy fertilizer	100 bushels	*	USDA ARMS Survey 1998
Soy manure surface	100 bushels	*	USDA ARMS Survey 1998
Soy manure inject	100 bushels	*	USDA ARMS Survey 1998
Other fertilizer	$ (value of production)	*	USDA ARMS Survey 1998
Other manure surface	$ (value of production)	*	USDA ARMS Survey 1998
Other manure inject	$ (value of production)	*	USDA ARMS Survey 1998
Hogs	cwt	*	USDA ARMS Survey 1998

* Estimated mean value varies by region and size of operation.

Appendix table A-2. Output price, P_{ir}

Outputs	Units	Value	Source
Corn (all)	$/100 bushels	284	NASS (average price 1997-99)
Soy (all)	$/100 bushels	700	NASS (average price 1997-99)
Other (all)	-	1	-
Hogs	$/cwt	46.92	NASS -(average price 1997-99)

Appendix table A-3. Input price, W_{jr}

Inputs	Units	Value	Source
Land	$/acre	68.2	NASS Agricultural Land Values Final Estimates 1998, Statistical Bulletin Number 957 (national average) (use 7% of land value as rental rate)
Capital	$	1	(by definition)
Feeder pigs	$/cwt	80.25	NASS (average price 1997-99)
Feed corn	$/100 bushels	284	NASS (average price 1997-99)
Feed soy	$/100 bushels	700	NASS (average price 1997-99)
Fertilizer N	$/lb.	0.185	Ribaudo et al., 2003

Appendix table A-4. Resource use, A_{ijr}

Input-output	Units	Value	Source
Land-corn	acres/100 bushels	*	USDA ARMS Survey 1998
Land-soy	acres/100 bushels	*	USDA ARMS Survey 1998
Land-other	acres/$	*	USDA ARMS Survey 1998
Capital-corn	$/100 bushels	49.3	Foreman, 2001
Capital-soy	$/100 bushels	127	Foreman, and Livezey, 2002
Capital-other	Share of value	0.17	Same share as corn
Capital-hogs	$/CWT.	*	USDA ARMS Survey 1998
Feed corn-hogs	100 bushels /CWT.	*	USDA ARMS Survey 1998
Feed soy-hogs	100 bushels /CWT.	*	USDA ARMS Survey 1998
Feeder pigs-hogs	CWT/CWT	*	USDA ARMS Survey 1998
Fertilizer-N-corn	lbs./ 100 bushels	80.0	Kellogg et al., 2000.
Fertilizer-N-soy	lbs./ 100 bushels	236.7	Kellogg et al., 2000.
Fertilizer-N-other	lbs./ $	0.282	Same rate as corn

* Estimated mean value varies by region and size of operation.

Estimate Calibrated Quadratic Cost Function

In step 2 we define quadratic total variable costs as $\frac{1}{2}\hat{Q}_{ir}X_{ir}^2$, where $\hat{Q}ir = (\hat{1}_{ir} + \hat{1}_{crop,r} + C_{ir}) / X0_{ir}$, $\hat{1}_{ir.}$, are the estimated dual values associated with equation 3 the calibration constraints, and $\hat{1}_{crop,r}$, are the estimated dual values associated with equation 4 the calibration constraints for each crop activity: crop Î {*corn, soybean, other*}. Since equation 4 applies only to crops, $\hat{Q}_{ir} = (\hat{l}_{ir} + C_{ir}) / X0_{ir}$ for *i=hogs*. The objective in step 2 is to maximize total net revenues:

$$\max_{x2_{ir}} \text{å}_r \text{å}_i P_{ir} X2_{ir} - \frac{1}{2}\hat{Q}_{ir} X2_{ir}^2 \tag{5}$$

subject to the resource constraints:

$$\text{å}_i A_{ijr} X2_{ir} \text{ £ } \text{å}_i A_{ijr} X0_{ir}, \text{ " } j, r \tag{6}$$

Solution of the non-linear optimization problem defined by equations 5 and 6 results in the initial output levels $X0_{ir}$.

Estimate Activity Levels for Policy Scenarios Using Calibrated Cost Function

Having characterized the farmer's non-linear optimization problem that results in the observed initial values, the final step is to impose policy constraints and compare solutions to the initial values. The policies we consider are the CAFO nitrogen application constraint and a hypothetical ammonia emission constraint. Farms can respond to policy constraints by adjusting input and output levels. Pit storage operations can vary the amount of land on which they inject versus surface-apply manure slurry in order to alter the ammonia emitted to the air and the nutrients available to plants. Lagoon operations can cover their lagoons to reduce air ammonia emissions. EQIP payments can enter the farmer's decision problem by reducing costs of abiding by the CAFO rules.[13]

First we incorporate into the optimization a manure transportation cost that depends on how the manure is stored and handled. Prior to implementation of the CAFO manure application rules, farmers had little incentive to transport manure off-farm, and few did. According to the 1998 survey, fewer than 2 percent of farms transported manure off-farm. The CAFO manure application rules require farmers to apply manure at a rate that plants can absorb. In response to the CAFO rules, farmers without adequate cropland will need to transport some manure off-farm (Ribaudo et al., 2003).

For the policy analysis, the farmer's objective is:

$$\max_{X3_{ir},\, COV_r} \sum_{r} \sum_{i} P3_{ir} X3_{ir} - \frac{1}{2} \hat{Q}_{ir} X3_{ir}^{2} - (1 - EQIP)MTC_r - COV_r \ast \kappa \times X3_{hogs,r} \tag{7}$$

where MTC_r is the cost of transporting manure off-farm, which is a function of technology choices that affect that nutrient availability to the crop—and consequently the amount of land on which the manure must be spread. Farms eligible for EQIP payments receive a share of the manure transportation costs and receive a per acre subsidy for land on which they apply manure at the agronomic rate. *EQIP* is defined as the share of manure transportation costs financed by EQIP. The per-acre EQIP subsidy is expressed as a per-unit subsidy and appears in the optimization as a higher price *P3*. The decision by lagoon farms to cover their lagoon is reflected in the binary choice variable COV_r (1 if covered, 0 otherwise). The cost of covering a lagoon is simply a cost κ per unit of hog output.

Manure transportation costs depend on the nutrient content of the manure (how it was stored), how it is applied (injected or spread), the availability of land on which to apply the manure, and what crops it is applied to. Estimates for the transportation costs per hundredweight of hog are based on a transportation cost model proposed by Fleming et al. (1998) (see Appendix table A-5 for details). Manure transportation costs equal the quantity of hogs used to produce manure transported off-farm, $hogs\_off_r$, multiplied by the manure transportation costs per hundredweight of hog. Manure transportation costs are distinguished for lagoon operations, which may or may not cover their lagoons:

$$MTC_r = hogs\_off_r(COV_r * T_{cover,r} + (1 - COV_r) * T_{uncov,r}), \tag{8}$$

and for pit storage operations which may inject (versus surface-apply) manure into some portion of the land on which manure is applied:

$$MTC_r = hogs\_off_r(INJ_r * T_{inject,r} + (1 - INJ_r) * T_{surf,r}), \tag{9}$$

where transportation costs per hundredweight of hog produced, $T_{e.r}$, depend on the manure storage and handling technology $e \in$ {covered, uncovered, surface-applied, injected}.

For lagoon operations, COV_r is a binary choice variable. For pit storage operations, INJ_r is the share of manure-applied cropland on which manure is injected:

$$INJ_r = \sum_{i \in mi} A3_{i,land,r} X3_{i,r} \Big/ \sum_{i \in m} A3_{i,land,r} X3_{ir}, \tag{10}$$

where m is the set of manure crop activities (corn, soybean and other crops, either spread or injected) and mi is the set of all cropping activities on which manure is injected.

The quantity of hogs that produce manure applied off-farm equals the total hogs produced minus the number of hogs required to produce the nitrogen from manure applied on-farm:

$$hogs\_off_r = X3_{hogs,r} - \left(manrate_r \sum_{i \in m} X3_{ir} A_{i,ferN,r} \left(\frac{COV_r}{NH_{cov}} + \frac{1 - COV_r}{NH_{uncov}} \right) \right) \tag{11}$$

Appendix table A-5. Manure off-farm transportation net costs by region and manure storage and handling technology, T_{re}

Manure storage/handling technology	Eastern Cornbelt	Western Cornbelt	Mid-Atlantic	South and West
	Dollars/cwt of hogs			
Lagoon				
Uncover	1.33	1.36	2.01	2.15
Cover	5.32	5.38	6.57	6.83
Pit				
Surface	1.20	1.25	2.29	2.53
Inject	1.61	1.66	2.82	3.08

Source: Estimated. Base manure handling costs from Fleming et al. 1998. Unit mile cost from USDA, NRCS, 2003 *Costs Associated with Development and Implementation of Comprehensive Nutrient Management Plans. Lagoon cover costs from Massey, et al. Agronomic and economic impacts of lagoon based swine operations complying with the proposed EPA zero discharge rule.*

The number of hogs required to produce the nitrogen from manure applied on-farm equals the manure nitrogen used on-farm divided by the manure nitrogen available to crops per hundredweight of hogs, NH_e (which depends on the cover technology). The manure nitrogen used on-farm equals the pounds of manure nitrogen applied on farm if it were applied at an agronomic rate $\sum_{i,m} X3_{ir} A_{i,fertN,r}$ (the rate at which chemical fertilizers are applied) multiplied by the factor $manrate_r$. From the survey we observe the average rate at which manure is applied to receiving land, but we do not know the rate applied to individual crops. Consequently, we assume that farmers apply manure at the same observed factor, $Manrate_r$, above the agronomic rate for all crops.

There is an analogous equation for pit storage operations.

$$hogs\_off_r = X3_{hogs,r} - \left(manrate_r \sum_{i \in m} X3_{ir} A_{i,fertN,r} \left(\frac{INJ_r}{NH_{inject}} + \frac{1 - INJ_r}{NH_{surf}} \right) \right) \qquad (12)$$

Policy 1: Nitrogen application constraint. CAFO rules require a nutrient management plan that requires growers to apply manure nitrogen at or below the rate at which plants can absorb (the agronomic rate). This policy is imposed by constraining $manrate_r$ to be less than or equal to 1.

Policy 2: EQIP payments. The effect of EQIP payments can be modeled by adjusting the share of off-farm manure transportation costs borne by EQIP and by adjusting the per-unit subsidy for crops produced in accordance with CAFO application guidelines.

Policy 3: Ammonia nitrogen emission constraint. Hypothetical ammonia emissions regulations are modeled by imposing a limit, *Amlimit*, on the quantity of nitrogen from ammonia per-unit of hog produced. Nitrogen emissions per unit of hog produced, AmN_e, depend on manure storage and handling technologies. The ammonia emission constraint is:

$$COV_r * AmN_{Cover} + (1\ COV_r)* AmN_{Un\,\mathrm{cov}\,er}\ \pounds\ \text{Amlimit} \qquad (13)$$

for lagoon operations and:

$$INJ_r * AmN_{Inject} + (1\ INJ_r) * AmN_{Surface}\ \pounds\ \text{Amlimit} \qquad (14)$$

for pit storage operations. Note that the ammonia emission constraint does not depend on the quantity of manure transported off-farm. The application method (spread/inject) is assumed to be the same on-farm and off-farm.

Appendix table A-6. Nitrogen available to crops and nitrogen ammonia emissions by manure storage and handling technology

Manure storage/ handling techology	Soil nitrogen available to plants, $Npercwt_e$	Air ammonia emissions from house and storage	Air ammonia emissions from land application	Total air ammonia emissions, AmN_e
	Lbs/cwt			
Lagoon				
Uncover	1.53	7.21	0.42	7.62
Cover	5.07	2.69	1.39	4.08
Pit				
Surface	4.83	3.00	1.32	4.32
Inject	5.95	3.00	0.20	3.20

Source: US EPA *National Emission Inventory--Ammonia Emission from Animal Husbandry Operations*, 2004.

Appendix table A-7. EQIP payments per unit of output by crop and region

Crop	Uni	Eastern Cornbelt	Western Cornbelt	Mid-Atlantic	South and West
Corn	$/100 bu	8.87	8.28	53.00	49.70
Soybean	$/100 bu	27.44	24.44	85.62	86.92
Other	Share of value	0.05	0.11	0.13	0.17

Source: Estimated using EQIP program data, Farm Service Agency, USDA.

APPENDIX B. USMP MODEL

The USMP model accounts for production of major crops (corn, soybeans, sorghum, oats, barley, wheat, cotton, rice, hay, and silage) and confined animals (beef, dairy, swine, and poultry), comprising approximately 75 percent of crop production and more than 90 percent of livestock and poultry production in the United States. USMP is a comparative-static, spatial, and market equilibrium model that incorporates agricultural commodity, supply, demand, environmental impacts, and policy measures. The model permits agricultural sectors to adjust to nutrient standards for air and water by substituting across space, production activities, and cropping and tillage practices with varying input requirements.

Crop and animal production choices are linked to edge-of-field environmental variables using the Environmental Policy Integrated Climate Model (EPIC), which uses a daily time step to simulate weather, hydrology, soil temperature, erosion-sedimentation, nutrient cycling, tillage, crop management and growth, and pesticide movements to the field's edge (Mitchell et al., 1998). The transport of nutrients, pesticides, and sediment across the landscape is calibrated to USGS estimates of regional pollutant loads (Smith, Schwartz, and Alexander, 1997).

Estimates of CAFO and AFO spreading practices on hog operations taken from Ribaudo, Gollehon, and Agapoff (2003) allow us to account for prior land application of manure in the simulations. Accordingly, CAFOs are assumed to spread manure on the nearest 155 acres and the smaller AFOs are assumed to spread manure on the nearest 90 acres. While these numbers are not necessarily representative of the range of production conditions across the Nation, we feel that these are reasonable for initial estimates of the environmental effects of excess manure application at the Farm Production Region scale. The above levels provide a lower bound on the estimated costs from meeting nutrient standards since many livestock facilities have little or no land on which to spread manure (Kaplan, Johansson, and Peters, 2004). Given the acres currently receiving manure nutrients, we calculate the quantity of manure nutrients in excess of the crop requirements on those acres. These excess nutrients are subject to leaching, runoff, and volatilization, similar to commercial fertilizers.

Manure transportation costs are determined using the Fleming et al. (1998) formulation in conjunction with regional and species-specific cost coefficients from the literature (Borton et al., 1995; Pease et al., 2001). The costs to develop a nutrient management plan, and to test periodically for manure nutrient composition and soil nutrient content are also included using USDA estimates (USDA, NRCS, 2003). Current market values for commercial nitrogen and phosphorus are used to calculate the savings from substituting manure nutrients for commercial fertilizers. The costs of using manure nutrients (testing, transporting, and applying) as fertilizers are covered by the livestock sectors. The savings in forgone commercial nutrient purchases by cropping enterprises are included in the returns to crop production.

APPENDIX C. MODELING MANURE MANAGEMENT IN THE CHESAPEAKE BAY WATERSHED

Our model is designed to minimize the total regional costs of manure management, transport, and application for use on agricultural lands in the Chesapeake Bay Watershed (CBW), given the existing structure and scale of the animal industry, and the current manure storage technology. The regional specification captures the element of competition for a limited land base by modeling access to spreadable land, requiring adequate area for land application of manure produced, and computing the associated hauling costs. Technologies that limit ammonia-N emissions alter regional competition by changing the costs and manure nutrient content across manure systems and animal types. Explicit modeling of competition for land on which to spread manure is a central feature of the regional model that is not captured in existing farm-level models.

The model was developed to: (1) provide a mechanism to track manure and related nutrient flows within the basin, from farm to site application and use, (2) compute the regional costs of land-applying manure, given the manure movement dictated by the nutrient uptake, and (3) provide a framework for evaluating alternative technologies that limit ammonia-N emissions, given land-application rates to meet a water-quality standard.

The county serves as the primary modeling unit for the regional model. The county-level specification provides consistency with Census of Agriculture data and other data, while permitting differentiation of institutions and regulatory conditions across county and State political boundaries within the watershed. County and local data are used to capture heterogeneity in technologies and land-quality conditions across the region, though our model may not represent the conditions on any particular farm.

The model is designed to minimize the regional cost of applied manure, subject to total manure produced and the land available for manure applications. Total regional costs of applied manure include transporting the manure, applying it to the land, implementing a nutrient management plan, implementing ammonia-reducing technology, based on 1997 production numbers. The model allocates manure flows between source and destination counties in the watershed to minimize the costs of hauling and applying manure, selected treatment costs, and costs of nutrient management plan development, given constraints on ammonia emissions and nutrient application rates. For a more detailed description of the water-based model, see Appendix 4-A in Ribaudo et al. (2003) or the technical documentation in Aillery and Gollehon (2004).

Including Air Emissions in the Modeling Framework

The regional modeling framework developed for manure management and water-quality policy analysis was extended to consider air emission measures. Air emissions were incorporated into the modeling framework by (1) adjusting the manure-nutrient content, (2) including treatment costs, and (3) calculating levels of ammonia emissions.

Changes in manure N content were calculated based on manure-nutrient adjustments by species, type of manure-handling system, and ammonia reduction measures. Changes in the N content of manure impact both the level of manure-N excess that must be transferred off confined animal farms and the rate of applied manure under an N-standard. Thus, implementation of policies to address air emissions issues will affect costs to the animal sector of meeting water-quality regulations.

The costs of emission control policies reflect the individual treatment costs for the three ammonia-reducing technologies considered-alum, incorporation into the soil, and lagoon covers-weighted by the share of acreage by species and manure system type, and use shares by treatment. Emissions were calculated by treatment scenario at both the storage facility (pre-haul) and field levels, for both regulated and non-regulated farms. Facility emissions are exogenous to the model, based on total manure production allocated across manure storage systems. Field emissions on regulated farms are calculated based on endogenously derived values for total land-applied manure (net industrial uses and that exceeding land capacity) and rate of applied manure in receiving counties. Field emissions on non-regulated farms were calculated from that portion of manure not explicitly addressed in the model optimization.

Model Data

Three primary data sources form the basis of the CBW model data set: the 1997 Census of Agriculture and the National Land Cover Dataset from USGS form the basic model structure and the National Emission Inventory from EPA is the source of the ammonia-N emission values. Farm-level Census data were used to generate county-level measures of animal operations and animal-units, total manure production, surplus recoverable manure, manure-nutrient content, and potential assimilative capacity of the land for applied manure nutrients. The National Land Cover Dataset was used to define the spatial pattern of land available for manure spreading and to simulate the spatial distribution of livestock operations (Ribaudo et al., 2003).

Model data on ammonia-N emissions were developed from system loss values presented in EPA's National Emission Inventory (NEI). For each manure-handling system, ammonia-N loss and retention are reported for animal confinement area, manure storage area, and land application area, based on a mass-balance approach. Starting from an excreted level of nitrogen in the manure, each unit of nitrogen will be either lost to the atmosphere or applied to the land for crop use.[14] Ammonia losses were aggregated for CBW model use based on losses from animal confinement and manure storage areas (termed "facility" losses) and subsequent losses during field application (termed "field losses"). The coefficients for ammonia-N losses were then derived at the facility and field levels, with losses expressed as a share of manure nitrogen available to the crop (and not as a share of excreted levels).

The shares of ammonia-N losses were then mapped to recoverable manure nitrogen available for plant use from Kellogg et al., (2000) to estimate the ammonia-N losses at each stage of the manure handling system.[15] Excreted manure nitrogen levels were derived from this mapping procedure for 1997 animal stocks in the CBW. For scenarios evaluating alternative technologies to reduce ammonia-N emissions, the process operated in reverse. From the calculated excreted nitrogen quantities, revised facility and field losses were subtracted to estimate a revised level of nitrogen available for crop use relative to the values in Kellogg et al., which constitute the core of the model data.

Production Cost Data

The NRCS Cost and Capabilities Assessment was the primary source of cost data for nutrient management plan components (USDA, NRCS, 2003). Manure hauling and application charges were based on published literature (Pease et. al., 2001; Fleming et. al., 1998), supplemented with data from the NRCS Cost and Capabilities Assessment. Transportation charges reflect a base rate per wet ton (loading/unloading and application) and hauling cost per ton-mile, by hauling mode and distance interval. Application costs are incorporated within hauling charges for lagoon and slurry systems; an additional charge was included for dry manure application. Per-acre costs of manure incorporation/injection were based on an Iowa State Farm Survey (2001). The baseline values assume that 40 percent of cropland acres currently incorporate manure, derived from information obtained in the ARMS hog and dairy surveys.

Chemical fertilizer costs were based on reported 1997 NASS prices, based on representative fertilizer products for the northeast States (USDA, NASS 2001). Cost-savings for reduced field application costs (under an N-standard) of $5 per acre were from Fleming, 1998. Annual costs associated with improved manure management practices to reduce ammonia-N emissions were: alum–$26.77 per poultry animal unit (AU) plus the additional hauling costs from adding an additional 10 percent to the weight of the litter; lagoon covers–$0.72 per AU for biofilter covers and $5.76 per AU for impervious covers; and incorporation/injection–$6.00 per acre. For a detailed description of the cost data see Appendix 4-A in Ribaudo et al. (2003) or the technical documentation in Aillery and Gollehon (2004).

For these systems, the share of N lost in each stage of the manure system was derived using a mass-balance approach based on manure management systems described by EPA (detailed in Section 2). Implementation of manure management practices to reduce ammonia-N emissions affects air emissions at different stages in the system. Alum affects the emissions from confinement structures while lagoon covers affect emissions from manure storage systems. These practices, which reduce ammonia emissions at the facility level prior to field application, actually increase volatilization during land application with surface application methods due to the manure's higher nitrogen content and expanded acreage requirements.

Incorporation/injection is a manure management practice that reduces ammonia emissions at the field level only. Field treatments can be used in combination with facility reduction practices or alone. In general, reducing the losses of nitrogen to the atmosphere increases the nitrogen level of manure available for crop use, and net reductions in emissions need to consider interactive effects from a broader systems perspective. Appendix table D-3 presents the model's assumptions regarding the changes in ammonia emissions and changes in the nitrogen level of the manure available for crop use. Appendix table D-4 presents examples of derived facility and field emissions using the coefficients in Appendix table D-3 for the major CBW manure systems in Appendix table D-2.

APPENDIX D. DATA AND COMPUTATION TABLES

Appendix table D1. Base representative systems and ammonia-N emissions for use in the CBW model, by animal type and manure system type

	Representative system		
Animal type	**Lagoon**	**Slurry**	**Dry/litter**
Dairy	Flush barn, surface applied	Pond storage, surface applied	Dry, solids system, surface applied
Feedlot beef	Flush barn, surface applied	Pond storage, surface applied	Dry, solids system, surface applied
Swine	Daily flush, surface irrigate	Deep pit, surface applied	Minor technology (used values for dry feedlot beef system)
Poultry	Not considered	Not considered	Broiler house, surface-applied litter
Facility emissions coefficients (share of N available to the crop)			
Dairy	4.242	0.637	0.309
Feedlot beef	4.242	0.637	0.309

Animal type	Lagoon	Slurry	Dry/litter
Swine	4.725	0.621	0.309
Poultry	n/a	n/a	0.417
Field emissions coefficients (share of N available to the crop)			
Dairy	0.282	0.209	0.0205
Feedlot beef	0.282	0.209	0.0205
Swine	0.274	0.274	0.0205
Poultry	n/a	n/a	0.333

Appendix table D-2. Manure ammonia-N production and losses for selected animal types and manure management systems for the CBW, baseline scenario

Animal type	Example manure system	Excreted nitrogen	Facility loss as ammonia-N	Field loss as ammonia-N	Nitrogen available for plant use
	Pounds per AU				
Dairy	Dry/Litter	99.2	20.25	13.43	65.5
Feedlot beef	Slurry	85.8	29.61	9.72	46.5
Swine	Lagoon	248.5	195.74	11.35	41.4
Poultry	Dry/Litter	421.8	100.5	80.26	241.0

Appendix table D-3. Ammonia-N emission and manure nitrogen changes with evaluated manure system improvements for use in the CBW model, by animal type and manure system type

Animal type	Item	Lagoon	Slurry	Dry/ litter
Manure Management System Change: Alum to poultry litter (coefficients expressed as a share of N available to the crop)				
Poultry	Facility emissions	n/a	n/a	-0.679
Poultry	Field emissions	n/a	n/a	0.213
Poultry	Applied manure nitrogen	n/a	n/a	0.2125
Manure Management System Change: Biofilter Lagoon Cover (coefficients expressed as a share of N available to the crop)				
Dairy, feedlot beef, and swine	Facility emissions	-0.264	n/a	n/a
Dairy, feedlot beef, and swine	Field emissions	0.979	n/a	n/a
Dairy, feedlot beef, and swine	Applied manure nitrogen	0.979	n/a	n/a

Appendix table D-3. (Continued)

Animal type	Item	Lagoon	Slurry	Dry/ litter
Manure Management System Change: Impervious Lagoon Cover (coefficients expressed as a share of N available to the crop)				
Dairy, feedlot beef, and swine	Facility emissions	-0.627	n/a	n/a
Dairy, feedlot beef, and swine	Field emissions	2.326	n/a	n/a
Dairy, feedlot beef, and swine	Applied manure nitrogen	2.326	n/a	n/a
Manure Management System Change: Incorporate/Inject Manure (coefficients expressed as a share of N available to the crop)				
Dairy and feedlot beef	Facility emissions	0.0	0.0	0.0
Dairy and feedlot beef	Field emissions	-0.75	-.80	-0.18
Dairy and feedlot beef	Applied manure nitrogen	0.212	0.233	0.06
Swine	Facility emissions	0.0	0.0	0.0
Swine	Field emissions	n/a	n/a	-0.18
Swine	Applied manure nitrogen	n/a	n/a	0.06

Appendix table D-4. Manure ammonia-N production and losses for selected animal types and manure management systems for the CBW, Ammonia-N reduction scenarios

Animal type	Example manure system	Scenario	Excreted nitrogen	Facility loss as ammonia-N	Field loss as ammonia-N	Nitrogen available for plant use
	Pounds per AU					
Poultry	Dry/litter	Alum	421.8	32.26	97.27	292.2
Swine	Lagoon	Impervious cover	248.5	73.01	37.74	152.2
Dairy	Dry/litter	Incorporation	99.2	20.25	11.02	68.0

References

Abt Associates. (2000). *Air Quality Impacts of Livestock Waste*. Report prepared for U.S. EPA Office of Policy. Bethesda, MD.

Aillery, Gollehon, M. N. & Breneman, V. (2005), "Technical Documentation of the Regional Manure Management Model for the Chesapeake Bay Watershed," Technical Bulletin No. 1913, Economic Research Service, U.S. Department of Agriculture, March, www.ers.usda.gov/publications/tb1913/

Arogo, J., Westerman, P. W., Heber, A. J., Robarge, W. P. & Classen, J. J. (2002). *Ammonia Emissions from Animal Feeding Operations*. National Center for Manure and Animal Waste Management, North Carolina State University, Raleigh, NC.

Baker, L. A., Hope, D., Xu, Y. & Edmonds, J. (2001). "Multicompartment Ecosystem Mass Balances as a Tool for Understanding and Managing the Biogeochemical cycles of Human Ecosystems," *The Scientific World*, *1*(S2), 802-808.

Baumol, W. & Oates, W. (1988). *The Theory of Environmental Policy*. Cambridge University Press, Cambridge, MA.

Borton, L. R., Rotz, C. A., Person, H. L., Harrigan, T. M. & Bickert, W. G. (1995). "Simulation to Evaluate Dairy Manure Systems." *Applied Engineering in Agriculture*, *11*, 301-310.

Bull, K. R. & Sutton, M. A. (1998). "Critical Loads and the Relevance of Ammonia to an Effects-Based Nitrogen Protocol," *Atmospheric Environment*, *32*(3), 565-572.

Carmichael, J. J. & Strzepek, K. M. (2000). "A Multi-Organic-Pollutant Simulation/Optimization Model of Industrial and Municipal Wastewater Loading to a Riverine Environment," *Water Resource Research*, *36*(5), 1325-1332.

Cody, C. (2003. "Farmers told to stop spreading litter." *Arkansas DemocratGazette*, April 29.

Conner, J. D., Perry, G. M. & Adams, R. M. (1995). "Cost-Effective Abatement of Multiple Production Externalities," *Water Resources Research*, *31*(7), 1789-96.

Council for Agricultural Science and Technology (CAST). (2002). "Animal Diet Modification to Decrease the Potential for Nitrogen and Phosphorus Pollution." Issue Paper 21 (July).

De Lucia, R., McBean, E. & Harrington, J. (1978). "A Water Quality Planning Model with Multiple Time, Pollutant and Source Capabilities," *Water Resources Research*, *14*(1), 9-14.

Feinerman, E., Bosch, D. & Pease, J. (2004). "Manure Application and Nutrient Standards," *American Journal of Agricultural Economics*, *86*(1), 14-25.

Fleming, R., Babcock, B. & Wang, E. (1998). "Resource or Waste? The Economics of Swine Manure Storage and Management" *Review of Agricultural Economics*, *20*(1), 96-113.

Follett, R. F. & Hatfield, J. L. (2001). *Nitrogen in the Environment*. Elsevier Press, Amsterdam, The Netherlands. 329 pp.

Foreman, L. F. (2001). *Characteristics and Production Costs of U. S. Corn Farms*. SB-974, U. S. Department of Agriculture, Economic Research Service, Aug.

Foreman, L. F. & Livezey, J. (2002). *Characteristics and Production Costs of U. S. Soybean Farms*. SB-974-4, U. S. Department of Agriculture, Economic Research Service, March.

Gollehon, N., M. Caswell, M. Ribaudo, R. Kellogg, C. Lander & D. Letson. (2001). *Confined Animal Production and Manure Nutrients*. AIB771, U. S. Department of Agriculture, Economic Research Service, June.

Hatfield, J. (2003). *Air Quality Primer*. Draft report.

Helfand, G. E. (1994). "Pollution Prevention as Public Policy, An Assessment," *Contemporary Economic Policy*, *12*(Oct.), 104-113.

Hohmann, H. 1994. "Cross-Media Pollution and International Law," *Natural Resources Journal* 34(Summer), 535-561.

House, R. M., McDowell, H., Peters, M. & Heimlich, R. (1999). "Agriculture Sector Resource and Environmental Policy Analysis, An Economic and Biophysical Approach," in *Environmental Statistics, Analyzing Data for Environmental Policy*, John Wiley and Sons, New York, NY.

Howitt, R. E. (1995). "Positive Mathematical Programming," *American Journal of Agricultural Economics*, *77*(2), 329-42.

Innes, Robert. (2000). "The Economics of Livestock Waste and Its Regulation," *American Journal of Agricultural Economics*, *82*(1), 97-117.

Jacobson, L. D., Moon, R., Bicudo, J., Janni, K., Noll, S., Shurson, G., Zhu, J., Schmidt, D., McGinley, C., Goodrich, P., Nicolai, R., Clanton, C., Davis, K. Brosseau, L., Bruns, J., Pijoan, C. & Blaha, T. (1999). *Generic Environmental Impact Statement on Animal Agriculture; A Summary of the Literature Related to air quality and Odor (H)*. Report prepared for Minnesota Environmental Quality Board, Minnesota Dept. of Agriculture.

Johansson, R. C. & Kaplan, J. D. (2004). "A Carrot and Stick Approach to Environmental Improvement, Marrying Agri-Environmental Payments and Water Quality Regulations," *Agriculture and Resource Economics Review*, *31*(1), 91-104.

Kaplan, J. D., Johansson, R. C. & Peters, M. A. (2004). "The Manure Hits the Land, Economic and Environmental Implications When Land Application of Nutrients is Constrained," *American Journal of Agricultural Economics* 86(3), 688-700.

Kellogg, R. L., C. H. Lander, D. C. Moffitt & N. Gollehon. (2000). *Manure Nutrients Relative to the Capacity of Cropland and Pastureland to Assimilate Nutrients: Spatial and Temporal Trends for the United States.* Natural Resources Conservation Service and Economic Research Service, USDA.

Lakshminarayan, P. G., Johnson, S. R. & Bouzaher, A. (1995). "A Multi-objective Approach to Integrating Agricultural Economic and Environmental Policies," *Journal of Environmental Management*, *45*, 365-378.

Lee, A. I. & Alm, J. (2004). "The Clean Air Act Amendments and Firm Investment in Pollution Abatement Equipment," *Land Economics*, *80*(3), 433-447.

Lence, B. (1991). "Weighted Sum Transferable Discharge Permit Programs for Control of Multiple Pollutants," *Water Resources Research*, *27*(12), 3019-3027.

Lipsey, R. G. & Lancaster, K. (1956). "The General Theory of Second Best," *Review of Economic Studies*, *24*(1), 11-32.

Maryland Department of Natural Resources. (2004). "Governor Ehrlich's Chesapeake Bay Watershed Restoration Fund." Press release for Chesapeake Bay Watershed Restoration Fund. www.dnr.state.md.us/dnrnews/pressrelease2004/talking_points.pdf#search='chesapeake%20bay%20watershed%20restoration%20fund'

Massey, R., Zulovich, J., Lory, J. & Millmier, A. (2004). "*Agronomic and Economic Impacts of Lagoon-Based Swine Operations Complying With the Proposed EPA Zero Discharge Rule*." mag/crops/massey/present/discharge/sld001. htm.

Metcalfe, S. E., Derwent, R. G., Whyatt, J. D. & Dyke, H. (1998). "Nitrogen Deposition and Strategies for the Control of Acidification and Eutrophication Across Great Britain," *Water, Air & Soil Pollution*, 107, 121-145.

Michaelis, P. (1992). "Global Warming: Efficient Policies in the Case of Multiple Pollutants," *Environmental and Resource Economics*, *2*(1), 61-77.

Mitchell, G., Griggs, R. H., Benson, V. & Williams, J. (1998). *Environmental Policy Integrated Climate Model.* Texas A&M University. www.brc.tamus.edu/epic/.

Moore, P. & Meisinger, J. (2003). "Air Quality Issues, Poultry". Working paper in J. Hatfield (ed.), *Air Quality Primer*. Draft report.

Moore, P. A., Jr., Daniel, T. C. & Edwards, D. R. (2000). "Reducing phosphorus runoff and inhibiting ammonia loss from manure with aluminum sulfate," *Journal of Environmental Quality* 29(1), 37-49.

Nagurney, A., Kanwalroop, K. & Stranlund, J. K. (1997). "General MultiProduct, Multi-Pollutant Market Pollution Permit Model: A Variational Inequality Approach," *Energy Economics*, *19*(1), 57-76.

National Research Council. (2003). *Air Emissions from Animal Feeding Operations: Current Knowledge, Future Needs.* Ad Hoc Committee on Air Emissions from Animal Feeding Operations, Committee on Animal Nutrition, National Research Council, National Academy Press, Washington, DC.

Oenema, O., Bannink, A., Sommer, S. G. & Velthof, G. L. (2001). "Gaseous Nitrogen Emissions from Livestock Farming Systems," in Follett, R. F. and J. L. Hatfied (eds.), *Nitrogen in the Environment: Sources, Problems and Management*. Elsevier Press, Shannon, Ireland.

Patton, M. & Seidl, A. (1999). *Matrix of State Level Policies for Animal Feeding Operations*. Agricultural and Resource Policy Report, Dept. of Agricultural and Resource Economics, Colorado State Univ., Ft. Collins, CO.

Pease, J., Pelletier, B. A. & Kenyon, D. (2001). "*Poultry Litter Transport Alternatives for Land Application in Virginia*." Selected paper presented at the Annual Southern Agricultural Economics Association meetings, Ft. Worth, TX, Jan. 29-31.

Preston, S. D. & Brakebill, J. W. (1999). *Application of Spatially Referenced Regression Modeling for the Evaluation of Total Nitrogen Loading in the Chesapeake Bay Watershed.* USGS Water-Resources Investigations Report 99-4054, U. S. Geological Survey, Reston, VA.

Randhir, T. O. & Lee, J. G. (1997). "Economic and Water Quality Impacts of Reducing Nitrogen and Pesticide Use in Agriculture," *Agricultural and Resource Economics Review*, *26*(1), 39-51.

Resources for the Future. (1996). "Cross-Media Pollution and the Chesapeake Bay," *Resources*, *124* (Summer).

Ribaudo, M., Gollehon, N., Aillery, M., Kaplan, J., Johansson, R., Agapoff, J., Christensen, L., Breneman, V. & Peters, M. (2003). *Manure Management for Water Quality: Costs to Animal Feeding Operations of Applying Nutrients to Land.* AER-824, U. S. Department of Agriculture, Economic Research Service.

Ribaudo, M., Catteneo, A. & Agapoff, J. (2004). "Cost of Meeting Manure Nutrient Application Standards in Hog Production, The Roles of EQIP and Fertilizer Offsets," *Review of Agricultural Economics*, *26*(4), 430-444.

Ribaudo, M. O., Horan, R. D. & Smith, M. E. (1999). *Economics of Water Quality Protection from Nonpoint Sources: Theory and Practice.* AER782, U. S. Department of Agriculture, Economic Research Service.

Ribaudo, M., Gollehon, N. & Agapoff, J. (2003). "Land Application of Manure by Animal Feeding Operations: Is More Land Needed?" *Journal of Soil and Water Conservation*, *58*, (1), 30-38.

Rohrer, W. (2004). "Delaware Nutrient Relocation Program." Presentation at Finding Solutions to Excess Nutrients in Animal Manure and Poultry Litter Conference, Beltsville, MD, Nov. 30.

Röhm, O. & Dabbert, S. (2003). "Integrating Agri-Environmental Programs into Regional Production Models, An Extension of Positive Mathematical Programming," *American Journal of Agricultural Economics*, *85*(1), 254-265.

Roos, K. F., Moser, M. A. & Martin, A. G. (1999). "*Agstar Charter Farm Program: Experience with Five Floating Lagoon Covers.*" Paper presented at the Fourth Biomass Conference of the Americas, Oakland, CA, Aug. 29-Sept. 2.

Smith, R. A., Schwarz, G. E. & Alexander, R. B. (1997). "Regional interpretation of water-quality monitoring data," *Water Resources Research*, *33*, 2781-2798.

Sweeten, J. (2001). "*Animal Production and Air Quality.*" Paper presented at Agricultural Outlook Forum (2001), Alexandria, VA, Feb.

Sweeten, J., Erickson, L., Woodford, P., Parnell, C., Kendall, T., Coleman, T., Flocchini, R., Reeder, C., Master, J., Hambleton, W., Bluhm, G. & Tristao, D. (2000). *Air Quality Research and Technology Transfer White Paper and Recommendations for Concentrated Animal Feeding Operations.* Report for USDA Agricultural Air Quality Task Force, Washington, DC.

Tietenberg, T. (1990). "Economic Instruments for Environmental Regulation," *Oxford Review of Economic Policy*, *6*(1), 17-33.

U. S. Department of Agriculture, Economic Research Service. (2005). "Confined Animal and Manure Data System." www.ers.usda.gov/data/manure/

U. S. Department of Agriculture, Economic Research Service. (2003). *Agricultural Resources and Environmental Indicators*, AH-722

U. S. Department of Agriculture, Natural Resources Conservation Service (NRCS). (2003. *Costs Associated with Development and Implementation of Comprehensive Nutrient Management Plans: Part I.*

U. S. Department of Agriculture, Natural Resources Conservation Service. (2005). "Draft Comprehensive Nutrient Management Planning Technical Guidance," *National Planning Procedures Handbook*, Subpart B, Part 600.51./www.nrcs.usda.gov/programs/afo/cnmp_guide_600.51.html.

U. S. Department of Agriculture, Natural Resources Conservation Service. (1999). "Nutrient Management." NRCS Conservation Practice Standard 590.

U. S. Department of Agriculture, World Agricultural Outlook Board (WAOB). (2003). "USDA Agricultural Baseline Projections to 2010," Staff Report WAOB-2003-1, U. S. Department of Agriculture, Office of the Chief Economist, Washington, DC.

U. S. Environmental Protection Agency. (2000). *Air Pollution Regulatory Assessment for Animal Feeding Operations*. Office of Air Quality Planning and Standards, Research Triangle Park, NC.

U. S. Environmental Protection Agency. (2004). *National Emission InventoryAmmonia Emissions from Animal Husbandry Operations*. Draft report www.epa.gov/ttnchie1/ap42/ch09/related/nh3inventorydraft_jan 2004.pdf.

U. S. Environmental Protection Agency. (2003). "National Pollutant Discharge Elimination System Permit Regulation and Effluent Limitations Guidelines and Standards for Concentrated Animal Feeding Operations: Final Rule," *Federal Register*, *69*(Feb.), 7175-7274.

U. S. Environmental Protection Agency. (2001). *Emissions from Animal Feeding Operations*. Draft report, Emissions Standards Division, Office of Air Quality and Planning and Standards, Research Triangle Park, NC.

U. S. Environmental Protection Agency. (1996). *Preventing Pollution Through Regulations: The Source Reduction Review Project: An Assessment*. EPA-742-R-96-001. Office of Prevention, Pesticides and Toxic Substances, Washington, DC.

U. S. Environmental Protection Agency. (2005). *Recommendations for Areas to be designated "Nonattainment" for the Fine Particle National Air Quality Standards*. Office of Air Quality Planning and Standards. http:// www.epa.gov/pmdesignations/documents/120/factsheet.htm.

U. S. Environmental Protection Agency, Office of Water. (1998). *National Water Quality Inventory: 1996 Report to Congress*. EPA841-R-97-008. Washington, DC.

U. S. Environmental Protection Agency. (2002). *State Compendium, Programs and Regulatory Activities Related to Animal Feeding Operations*. Tetra Tech, Fairfax, VA.

Weinberg, M. & Kling, K. L. (1996). "Uncoordinated Agricultural and Environmental Policy Making: An Application to Irrigated Agriculture in the West," *American Journal of Agricultural Economics*, *78* (1), 66-78.

Williams, M. (2004). "Progress of Identifying and Installing Potential Environmentally Superior Technologies per Agreements Between the North Carolina Attorney General, Smithfield Foods, Premium Standard Farms and Frontline Farmers. " Presentation at public meeting, July 14. North Carolina State University, Raleigh, NC. www.cals.ncsu.edu/waste_mgt/smithfield_projects/smithfieldsite.htm

Wilson, J. (2004). "AQMD Moves to Corral Cow Pollution," *Los Angeles Times*, June 17.

Wolfsohl, K. (2003). "A Big Stink: Small communities fight big farms." *Progressive Farmer*, (July), 28-29.

Zilberman, D., Ogishi, A. & Metcalfe, M. (2001). "Innovative Policies for Addressing Livestock Waste Problems," *White Paper Summaries*, National Center for Manure and Animal Waste Management, 51-53.

End Notes

[1] A recent court ruling has brought into question how many CAFOs will actually need NPDES permits. It may be less than 15,500

[2] Another option would have been to place a direct restriction on the entire farm. This would provide a different incentive to the farmer than the per-unit output restriction, and would likely result in a different outcome (Helfand, 1991). How different is an empirical question, but the direction of change would be the same.

[3] Here, though, we evaluate both nitrogen and phosphorus application standards. Because each region in the model is large, there is sufficient land to assimilate manure nitrogen in the baseline, meaning there is no overapplication. A nitrogen standard alone would result in no change in the model results.

[4] Manure generation is calculated according to Kellogg et al. (2000); crop nutrient demands by region are calculated using the Environmental Policy Integrated Climate Model (EPIC; Mitchell et al., 1998).

[5] We assume a crop producer willingness to accept manure of 30 percent, meaning that approximately 30 percent of available cropland in each region will utilize manure nutrients (USDA, ERS, 2003b). Alternative levels of manure utilization have been considered, but are not included here.

[6] Bioavailability of phosphorus refers to the amount of phosphorus in runoff that is available for aquatic and terrestrial plant growth.

[7] We elected to focus on the results for a 10-percent reduction in ammonia emissions (*Water, WaterAir10, and Air10*). At higher levels of constraints on air emissions, it is likely that

producers would consider many alternative technologies, which are not feasible to model at this point.

[8] Some of the nitrogen applied to cropland leaches to ground water, which is an important source of drinking water in many areas. The policies examined would have little impact on nitrogen leaching nationally (see figure 4-4).

[9] Increasing the nitrogen content of manure by adopting emission controls does not increase the acreage needed for land application when meeting a phosphorus standard. In fact, the increased nitrogen in the manure reduces the supplemental nitrogen usually required.

[10] Results here are very similar but not identical to those in Ribaudo et al., with the differences due to model and data improvements.

[11] The actual effect on water quality in the Bay will depend on the rate of nutrient loading from applied (and over-applied) manure, and the rate and location of nitrogen deposition from air emissions. The science behind these issues continues to evolve.

[12] The value of production by animal unit is determined by using USDA baseline projections for 2010 (USDA. 2003). These projections are in turn processed through the USMP model (see Appendix B, web only) to convert animal production into animal units and to account for production cycles. For our purposes, we use gross value minus variable costs per AU to represent the opportunity cost of decreasing production.

[13] We assume for this analysis that CAFOs do not receive EQIP payments.

[14] This assumption ignores direct discharge to water and accidental spills, which are believed to be non-significant.

[15] The values in Kellogg et al. were derived from the Census of Agriculture and are the basis for manure estimates in the model.

In: Manure Management
Editor: Sheila E. Corby

ISBN: 978-1-62948-874-5

Chapter 2

TRENDS AND DEVELOPMENTS IN HOG MANURE MANAGEMENT: 1998-2009*

Nigel Key, William D. McBride, Marc Ribaudo and Stacy Sneeringer

ABSTRACT

In the past decade, hog production has increasingly become consolidated, with larger operations producing a greater volume of hog manure on smaller areas. With less crop-land for spreading the manure, hog farmers may be compensating through more effective manure management. The authors use data from 1998 to 2009 collected in three national surveys of hog farmers. Over this period, structural changes in the hog sector altered how manure is stored and handled. Changes to the Clean Water Act, State regulations, and local conflicts over air quality also affected manure management decisions. The findings further suggest that environmental policy has influenced conservation-compatible manure management practices. The authors examine how the use of nutrient management plans and of practices such as controlled manure application rates vary with scale of production and how these practices changed over the study period. This chapter is an update of an earlier report, *Changes in Manure Management in the Hog Sector: 1998-2004*.

---

* This is an edited, reformatted and augmented version of the United States Department of Agriculture Publication, EIB-81, dated September 2011.

Keywords: hog production, manure management, structural change, environmental regulation

SUMMARY

What Is the Issue?

Over the last decade, U.S. hog industry production has shifted to fewer and larger operations that specialize in a single phase of hog production and the use of production contracts. Consolidation of operations has meant that an increasing volume of manure is often produced on farms with less cropland per animal for spreading the manure. A higher manure-to-cropland ratio has magnified the risk that manure nutrients (nitrogen, phosphorous, and potassium) and pathogens might flow into ground and surface water due to overapplication of manure on crops or leakage from manure storage facilities. These runoff contaminants can harm aquatic life and livestock and affect human drinking water. In addition, increased concentration of hogs per farm has led to conflicts with nearby residents or communities over odor and air quality. Legislative initiatives, such as the Clean Water Act and various State regulations, have implemented environmental policies to mitigate the risk of water pollution and reduce conflicts. Despite the pronounced shifts in hog industry structure and regulation, little information exists for assessing how these changes have affected manure management practices and environmental quality. Using data from three surveys of hog farmers, the authors shed light on these issues by examining how hog manure management practices vary with the scale of production and how practices have changed since 1998.

What Are the Major Findings?

- Hog farm operators altered their practices between 1998 and 2009 in ways that suggest intent to manage manure in a more environmentally sound manner. In 2009, operators were more likely to have comprehensive nutrient management plans and, accordingly, to have increased their efforts to apply manure at rates not exceeding the nutrient needs of the crop. They were more likely to have applied manure over a larger share of their cropland and to have increased their use of feed additives that reduce phosphorus in hog manure, increased the nutrient testing of manure, and removed excess manure from their premises. They were less likely to have added commercial fertilizer to crops receiving manure.

- Changes in manure management practices and outcomes from 1998 to 2009 are related to structural changes in hog production, particularly the increase in farm size and regional shifts of production. The changes include more use of pit/tanks and less of lagoons for manure storage; a decline in spreading solid manure, as well as in spreading liquid manure without incorporating it into the soil; and less intensive manure application by the largest farms.
- Environmental policies are likely behind some of the observed patterns of change in hog manure management between 1998 and 2009. For example, the relatively slow growth in production in the Southeast compared with the Heartland may be partly attributable to State regulations in North Carolina designed to reduce risks associated with manure lagoons. Nationally, greater use of comprehensive nutrient management plans, as well as a decline in intensity of manure applications by the largest operations, may be in response to Federal and State policies designed to reduce over application of manure nutrients. These changes suggest that larger hog operations are altering their manure management decisions in response to legal nutrient application constraints and that environmental policy is contributing to the adoption of conservation-compatible manure management practices.

How Was the Study Conducted?

This study uses information from surveys of U.S. hog producers conducted in 1998, 2004, and 2009 as part of USDA's annual Agricultural Resource Management Survey (ARMS). The detailed surveys cover a cross-section of U.S. hog operations and collect information on production costs, business arrangements, production facilities and practices, and farm operator and financial characteristics. The surveys also provide information about manure storage and handling, fertilizer use, manure application techniques, Environmental Quality Incentives Program (EQIP) payments, the use of comprehensive nutrient management plans, and manure application rates. The data allow the authors to document the current state of manure management and track producers' manure management practices during a period of rapid change in the hog industry. Data from the surveys are disaggregated by farm size according to the number of animal units produced (by 1,000 pounds of live animal weight). Because larger hogs produce more manure, animal units provide a consistent measure for comparing manure handling trends among farms with different levels of manure output.

INTRODUCTION

Over the past several decades, economic competition has driven substantial changes in the hog industry; production has shifted regionally and to larger operations that increasingly specialize in a single phase of hog production and use production contracts. In turn, the changing farm structure is altering manure practices as larger operations seek to manage nutrients on a smaller cropland base. At the same time, chemical fertilizer prices have increased markedly, providing an incentive for producers to adopt methods that conserve and better utilize manure nutrients.

Changes to the Clean Water Act, State regulations, and conflicts with nearby residents over odor and air quality are also influencing manure management decisions. Changes in the regulatory climate have been driven by increased environmental risk to air and water associated with larger livestock operations. Nutrients (nitrogen, phosphorous, potassium), ammonia, hydrogen sulfide, methane, pathogens, and odor can originate at production houses where animals are kept, at manure storage structures such as tanks and lagoons, or on land where manure is applied. The concentration of animals and manure into smaller areas increases the risk of environmental contamination, the nuisance potential of farms, and the overall challenge of managing manure.

We use data from recent surveys of hog farmers to examine how hog manure management practices vary with the scale of production and how practices changed between 1998 and 2009 (see box, "Data"). The findings provide information about the potential effects of recent policies and structural changes on manure management technologies and practices, the use of nutrient management plans, and manure application rates.

DATA

We analyzed information from surveys of U.S. hog producers conducted in 1998, 2004, and 2009 as part of USDA's annual Agricultural Resource Management Survey (ARMS). The detailed surveys cover a cross-section of U.S. hog operations each year and collect information on production costs, business arrangements, production facilities and practices, and farm operator and financial characteristics. The surveys also provide information about manure storage and handling, fertilizer use, manure application techniques, payments received under the Environmental Quality Incentives Program, the use of comprehensive nutrient management plans, and manure application rates.

The sample of hog farms was chosen from a list of operations maintained by USDA's National Agricultural Statistics Service. The target population of each survey was farms having 25 or more hogs at any time during the year. Farms with fewer than 25 hogs were screened out to exclude operations that raise hogs primarily for on-farm consumption and other noncommercial activities, such as youth projects. Each surveyed operation represents a number of similar farms in the population as indicated by the observation's expansion factor, or survey weight. The ARMS hog survey sampled operations in 22 States in 1998, 19 States in 2004, and 19 States in 2009. When the samples in each survey are expanded by their population weights, they represent more than 90 percent of the hog and pig inventory on U.S. farms in each survey year.

Estimates from the three surveys are comparable because of the consistent way in which the surveys were conducted and processed. Each survey had broad national coverage, represented the same target population, was conducted the same way (hand-enumerated) by the same organization, and collected much the same information in a similar format. More information about the ARMS and the hog surveys, including copies of the questionnaires, can be found at:www.ers.usda.gov/ Briefing/ARMS/.

Data from the surveys are disaggregated by farm size according to the number of animal units (1,000 pounds of live animal weight) produced. Because larger hogs produce more manure, animal units provide a consistent measure for comparing farms that produce hogs at different stages of the production cycle.

For example, farrow-to-finish and hog finishing operations produce much larger hogs, and thus more manure per animal, than do farrow-to-feeder pig and specialized nursery operations. The number of animal units on each farm is the steady-state live weight of sows, piglets, and other hogs, which is computed using average hog inventories and weights from the ARMS.

The environmental implications of hog production depend primarily on the manure management decisions of operations having at least 50 animal units. These operations accounted for 97 percent of hog sales in the sample frame in 1998 and 99 percent in both 2004 and 2009. For this reason, and to simplify the tabular presentations, statistics for operations with fewer than 50 animal units are not reported in most tables and figures.

Structural Change and Manure Management

Changes in manure management practices reflect changes in the scale, organization, geography, and methods of hog production. The production of hogs to be slaughtered for pork involves four phases: (1) breeding and gestation (breeding females and their maintenance during gestation); (2) farrowing (birth of pigs and care until weaning); (3) nursery (care immediately after weaning until about 30-80 pounds); and (4) finishing (feeding hogs from 30 to 80 pounds to a slaughter weight of 225-300 pounds). Hog producers are commonly classified according to the number of production phases conducted on the operation: (1) farrow-to-finish (all four phases); (2) farrow-to-feeder pig (phases 1, 2, and 3); (3) feeder pig-to-finish (phase 4); (4) wean-to-feeder pig (phase 3); (5) farrow-to-wean (phases 1 and 2); or (6) wean-to-finish (phases 3 and 4).

Historically, most U.S. hog production has occurred on farrow-to-finish operations located in areas with an abundant supply of corn. Hog farmers typically used corn produced on the farm as an inexpensive source of feed and applied hog manure as fertilizer on farm fields. Advancements in breeding and genetics, swine housing, nutrition, feeding, and farm management since the 1970s have allowed producers to become increasingly specialized in only one or two of the production phases. During the period of this analysis, hog farms using the farrow-to-finish approach declined from 49 percent in 1998 to 31 percent in 2004 and 23 percent by 2009 (table 1).

Table 1. Hog operation characteristics, 1998, 2004, and 2009

Characteristics for all farms	1998	2004	2009
All farms:			
Number of hog operations	61,971	40,940	24,394
Hog sales and contract removals (head per farm)	2,589	4,646	7,930
Hog inventory (animal units per farm)	194	294	546
Producer type: Farrow-to-finish (% of farms)	49	31	23
Producer type: Feeder pig-to-finish (% of farms)	31	40	46
Used a production contract (% of farms)	15	28	48
Value of production from hog enterprise (% total)	56	71	70
Acres of cropland per animal unit	2.14	1.41	0.86
Farms with no cropland (%)	8	19	19
Observations in sample	1,633	1,198	1,290

Note: A farm is defined as an operation having 25 or more hogs at any time during the year and includes independent hog producers and growers who produce hogs under contract. The sample covered operations in 22 States in 1998 and 19 States in 2004 and 2009. Animal units are defined as 1,000 pounds of live animal weight, and the inventory of animal units is based on an estimate of the average number of hogs and pigs on the operation in each year. Farrow-to-finish operations are those on which pigs are farrowed and then finished to a slaughter weight of 225-300 pounds. Feeder pig-to-finish operations are those on which feeder pigs are obtained from outside the operation, either purchased or placed under contract, and then finished to a slaughter weight of 225-300 pounds. Crop acres include land used for hay production, but exclude land used for pasture only.

Source: USDA, ERS, 1998, 2004, and 2009 Agricultural Resource Management Surveys.

Increases in the scale of production have accompanied changes in organization. Between 1998 and 2009, the number of hog operations fell by about 60 percent, and the average inventory grew from 2,589 to 7,930 head (table 1).

Over this period, the number of hog operations producing fewer than 300 animal units declined rapidly, resulting in a shift in production to larger operations (figures 1 and 2).

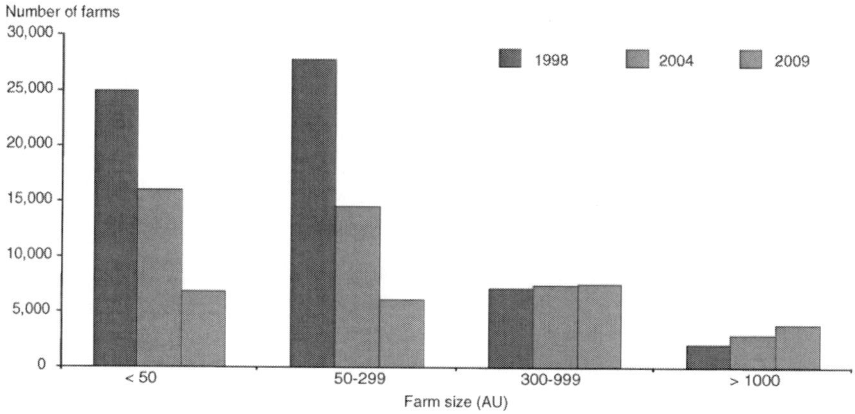

Note: Animal units (AU) are defined as 1,000 pounds of live animal weight, and the inventory of animal units is based on an estimate of the average number of hogs and pigs on the operation in each year.

Source: USDA, ERS, 1998, 2004, and 2009 Agricultural Resource Management Surveys.

Figure 1. Small hog operations declined in number and large operations grew between 1998 and 2009.

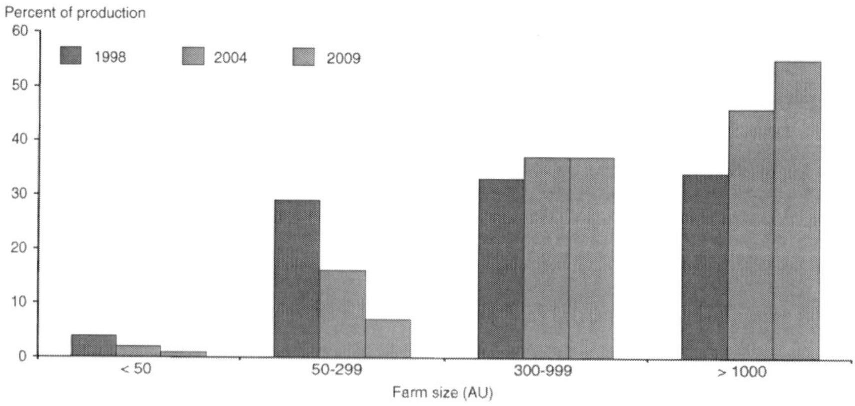

Note: Animal units (AU) are defined as 1,000 pounds of live animal weight, and the inventory of animal units is based on an estimate of the average number of hogs and pigs on the operation in each year.

Source: USDA, ERS, 1998, 2004, and 2009 Agricultural Resource Management Surveys.

Figure 2. Large operations produced an increasing share of output between 1998 and 2009.

The introduction of contract production arrangements has also played a significant role in the evolution of U.S. hog production. In contract production, a pig owner (the contractor) engages a producer (the grower) to take custody of the pigs and care for them in the producer's facilities in exchange for a predetermined formula for compensation. Contractors typically furnish inputs to growers, provide technical assistance, and collect the livestock to pass on for final processing or marketing. Packers or processors obtain hogs through marketing arrangements with the contractors (integrators) or directly from growers. Contract arrangements allow individual producers to specialize in one phase of production and increase their scale of operations.

Over the 11 years analyzed, the share of farms using a production contract more than tripled, and now almost half of all hog farms use contracts. An even larger share of output is now produced under contract, as the 48 percent of farms using production contracts in 2009 accounted for 71 percent of hog production. One issue with production contract arrangements is liability for managing the hog manure. Most contracts require growers to comply with all State, Federal, and local regulations in operating their facilities, and failure to comply can result in contract termination (Ogishi et al., 2003). Since contract growers are heavily invested in facilities, they are highly motivated to avoid liability.

Another structural change with implications for manure management is the increasing specialization by operators in the hog enterprise. Among hog farms, the contribution of hogs to the farm's total value of production increased from 56 to 70 percent over the study period, with all of the change occurring between 1998 and 2004 (table 1). Sources of hog feed also indicate increasing specialization in hog production – the share of feed consumed by hogs and produced on the same farm fell from about 50 percent in 1992 to below 20 percent by 2004 (Key and McBride, 2007). The increasingly specialized operations, with greater shares of production value from hogs and more hog feed from off-farm sources, often produce hogs under contract. In a contract arrangement, contractors deliver feed from off-farm sources to their growers. This allows individual growers to use their time and financial resources to increase the scale of their hog enterprise rather than to cultivate crop acreage for feed production. As a result, cropland per animal unit on hog operations has steadily declined from 2.14 acres in 1998 to 1.41 acres in 2004, and to 0.86 acres in 2009 (table 1). The share of operations with no cropland increased from 8 to 19 percent over this period.[1] This increase has created concerns about how to manage the large quantities of manure generated on hog operations with less available cropland per animal unit.[2]

Geographical shifts in hog production have accompanied structural and organizational changes in the industry. Historically, hog production was concentrated in the Heartland, particularly in Iowa and Illinois, where an abundant corn supply provided a cheap source of hog feed and sufficient acreage on which to spread hog manure. During the 1980s and 1990s, hog production grew dramatically in the Southeast, especially in North Carolina, driven mainly by the growth of large contract operations. Growth in the Southeast has posed the challenge of managing an increasing volume of hog manure in areas with a denser population and much less crop acreage for manure application than in the Heartland. In 2009, hog operations had about a quarter-acre of cropland per animal unit in the Southern regions compared with about 1 acre per animal unit in other regions (figure 3). Most hog producers in the Southeast treat and store manure in lagoon systems and apply manure to farm acreage with an irrigation system. In 2009, 90 percent of hogs produced in the South were on operations with a lagoon system compared with only 20 to 30 percent of hogs in other regions (figure 3). The lagoon-sprinkler manure system increases nitrogen volatilization and allows producers to apply hog manure on less acreage than other manure management systems.

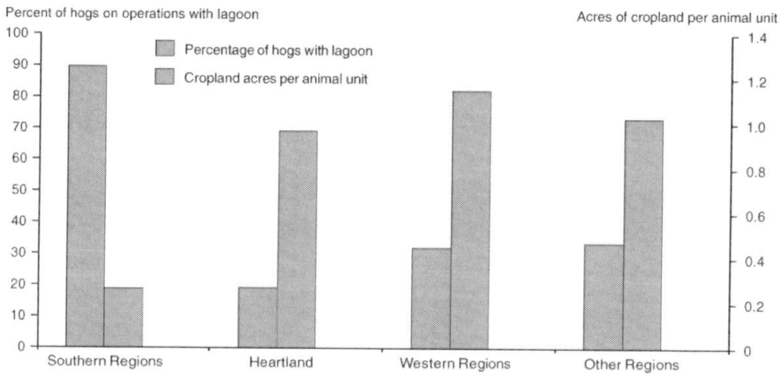

Note: Regions are ERS designations. Southern Regions include Southern Seaboard and Mississippi Portal; Heartland includes Heartland; Western Regions include Northern Great Plains and Prairie Gateway; Other Regions include Northern Crescent, Eastern Uplands, Fruitful Rim, and Basin and Range.

Map of regions can be found at: http://www.ers.usda.gov/Data/FarmandRelatedEmployment/DataFiles.htm

Source: USDA, ERS, 1998, 2004, and 2009 Agricultural Resource Management Surveys.

Figure 3. Lagoon use high and cropland availability low in Southern Regions, 2009.

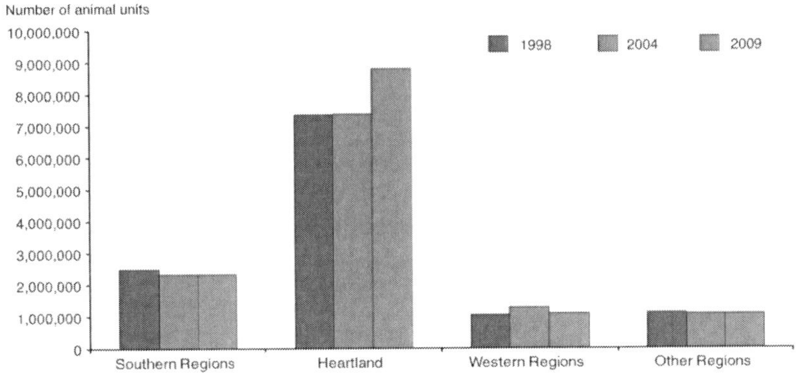

Note: Regions are ERS designations. Southern Regions include Southern Seaboard and Mississippi Portal; Heartland includes Heartland; Western Regions include Northern Great Plains and Prairie Gateway; Other Regions include Northern Crescent, Eastern Uplands, Fruitful Rim, and Basin and Range.

Map of regions can be found at: http://www.ers.usda.gov/Data/FarmandRelated Employment/DataFiles.htm/

Animal units (AU) are defined as 1,000 pounds of live animal weight, and the inventory of animal units is based on an estimate of the average number of hogs and pigs on the operation in each year.

Source: USDA, ERS, 1998, 2004, and 2009 Agricultural Resource Management Surveys.

Figure 4. Hog production has increased in Heartland region.

Between 1998 and 2009, the regional distribution shifted again: hog production increased in the Heartland and declined slightly in the South (figure 4).Production increased in Western States like Colorado and Utah, where low population density provides flexibility in managing animal manure. Production in the Heartland region now accounts for about two-thirds of U.S. hog output. Lack of growth in the South could be attributed partly to a 1997 moratorium on the construction of new and expanded hog operations in North Carolina (North Carolina General Assembly, 1997). The moratorium was enacted in response to environmental concerns about managing hog manure from increasingly larger operations.

Structural changes have coincided with substantial efficiency gains for hog farms, particularly on specialized hog-finishing operations. Most of these productivity gains were attributable to increases in the scale of production and technological innovation (Key, McBride, and Mosheim, 2008). Efficiency gains included higher feed productivity (output per unit of feed), which can reduce the amount of manure produced by hog operations and thus the amount of manure nutrients that must be disposed of.

Environmental Policy Background

Changes in environmental policies have also influenced manure management decisions. Many hog producers value the preservation of environmental quality and have found that profitable business practices are consistent with improved manure management practices. Nonetheless, concerns about water quality and air quality (primarily odor) have spurred local, State, and Federal action to mitigate environmental impacts of animal manure. The U.S. Environmental Protection Agency (U.S. EPA) revised Clean Water Act regulations in 2003 for controlling runoff of manure nutrients from the largest animal feeding operations (AFOs). These regulations were finalized in 2008. Clean Water Act regulations now require animal feeding operations designated as Concentrated Animal Feeding Operations (CAFOs) and discharging or proposing to discharge manure effluent to seek National Pollutant Discharge Elimination System (NPDES) permit coverage. Discharge from the production area and any land application areas under the control of the CAFO are included in the permit. CAFOs with permits must have a nutrient management plan that identifies site-specific practices to ensure agronomic use of nutrients.

Under the new regulations, CAFOs that are not required to have an NPDES permit, but that wish to claim the agricultural storm water exemption for runoff from fields receiving manure, must implement a nutrient management plan to minimize polluted runoff. If a waterway becomes polluted with animal waste from field runoff and a CAFO does not follow a nutrient management plan, the CAFO will be violating the Clean Water Act.

Enforcement of these updated Federal CAFO rules is largely devolved to individual States, which have adopted the new rules at various rates. In April 2010, the EPA reported that most States are expected to complete revisions to comply with the updated Federal rules by the end of 2010.

Atmospheric emissions of pollutants are regulated by the Clean Air Act (CAA). The CAA authorizes regulatory programs primarily for protecting human health. EPA recently initiated development of regulations for reducing fine particulates in the atmosphere (referred to as $PM_{2.5}$, for particles less than 2.5 microns in size). The Clean Air Act requires State, local, and tribal governments to identify areas not meeting national air quality standards for fine particles (one of the six criteria pollutants regulated under the act) (U.S. EPA, 2004). States with designated nonattainment areas were required to submit plans outlining how they would meet the standards by 2010. This regulation could affect animal operations because ammonia emitted from barns and manure storage facilities is a major precursor of fine particulates. Controlling ammonia from animal operations would be a likely priority in nonattainment areas with high concentrations of animals (U.S. EPA, 2000a). To gather data and develop generalized protocols, a 2006 EPA consent agreement encouraged livestock producers to volunteer for emissions monitoring in exchange for clemency for past air pollution violations (U.S. EPA, 2006). The National Air Emissions Monitoring Study (NAEMS) was concluded in 2010, and in early 2011 the EPA stated that the data were available for release (U.S. EPA, 2011).

Since 2010, a new Greenhouse Gas Reporting Rule has required owners and operators of facilities that contain manure management systems emitting at least 25,000 metric tons of greenhouse gases per year (expressed as carbon dioxide equivalents) to report emissions from all sources at the facility for which emission calculation methods are defined. EPA estimates that this rule will cover from about 85 to 95 of the Nation's largest livestock facilities (U.S. EPA, 2009b).[3]

Other legislation covering air pollution includes the Comprehensive Environmental Response Compensation and Liability Act (CERCLA) and the Emergency Planning and Community Right-to-Know Act (EPCRA). Both laws use information disclosure to increase the information available to governments and citizens about the sources and magnitude of chemical releases to the environment. CERCLA requires that facilities report to EPA when releasing more than a "reportable quantity" (e.g., 100 pounds in a 24-hour period) of a hazardous substance. EPCRA requires that a facility report to State and local authorities any releases reported under CERCLA. EPA is authorized to require long-term remedial action that significantly reduces threats to public health. Though CERCLA was originally focused on hazardous wastes from industrial plants, the increased size and consolidation of animal feeding operations may make their ammonia and hydrogen sulfide emissions subject to the notification provisions of the act. EPA has enforced the CERCLA and EPCRA reporting requirements of animal feeding operation releases of hazardous pollutants in two cases (Copeland, 2008).

However, due to controversy over the use of these laws for agricultural emissions (Copeland, 2008), the EPA issued a reporting exemption for air releases of hazardous substances from animal waste at farms in 2009 (U.S.

EPA, 2009a). The rule exempts all farms that release hazardous substances from animal waste to the air from reporting under CERCLA. Farms that do not meet the size definition of a CAFO (as defined by the Clean Water Act) are also exempt from reporting under EPCRA. However, farms that do meet the size definition of a CAFO are still required to submit the appropriate reports to State and local officials under EPCRA if they release hazardous substances to the air that meet or exceed their Reporting Quantity.

Odor arising from livestock operations has become a source of controversy in many agricultural communities, prompting numerous States to adopt some form of odor regulation (Redwine and Lacey, 2000; Sneeringer, 2010). These State policies may contain objective odor standards or require management methods such as separation distances, explicit odor plans, or "good neighbor practices." Several States include odor stipulations in their general permits for livestock operations (U.S. EPA, 2002b). To the extent that odor is correlated with levels of certain air emissions (particularly hydrogen sulfide and ammonia), State policies to limit such emissions may also help to control odor.

Most States have implemented regulations for controlling at least some of the environmental impacts of AFOs, including permits, licenses, and zoning requirements. To prevent large amounts of waste from being carried into coastal estuaries during major rain events such as hurricanes, North Carolina entered into a legal agreement with the State's largest swine producers to develop innovative waste management strategies that would replace uncovered lagoon and spray field systems for storing and treating waste from large hog operations (Williams, 2004; Bales et al, 2000). In 1997, the State issued a moratorium on the construction of new or expanded facilities in order to study and put in place a regulatory system to ensure that waste structures were sound, that waste application methods were adequate, and that viable waste utilization plans were in place. Iowa, Pennsylvania, Arkansas, and Kentucky are among States that have introduced rules for curbing water pollution, ammonia, and odor from AFOs (Patton and Seidl, 1999; U.S. EPA, 2002a,b).

Agricultural-residential conflicts at the rural-urban interface seem to be increasing as residential development expands further out into rural areas, while market conditions push farmers to intensify their production (Bergstrom and Centner, 1989; Centner, 2002; Duke and Malcolm, 2003; Jacobson et al., 2006). Conflicts that arise over environmental concerns are most prevalent for animal operations (Duke and Malcolm, 2003; Centner, 2002). Agricultural-residential proximity can result in citizen complaints to local authorities and actual and threatened lawsuits over perceived threats to health and environmental quality, even when no laws have been broken. Such actions may force farmers to modify their production practices. Adoption of "acceptable" nutrient management plans and practices is a way that farmers can demonstrate due care and possibly protect themselves from conflict over environmental quality (Centner, 2002).

To help defray the costs of meeting the regulations, producers can apply for financial assistance from the USDA's Environmental Quality Incentives Program (EQIP).[4] A farmer or entity may receive up to $300,000 for all EQIP contracts entered during a 6-year period to help develop and implement a nutrient management plan, construct appropriate animal and manure handling and storage facilities, or transfer and apply manure to land in an approved manner (Ribaudo, Cattaneo, and Agapoff, 2004).

Some major hog producers, such as Smithfield Foods, are taking the necessary steps for meeting voluntary environmental performance standards established by the International Organization for Standardization. ISO 14001 is the Environmental Management System (EMS) standard that measures how well companies manage those parts of their operation that affect the environment (Wall et al., 2001). The EMS includes such elements of hog production as sanitation, waste treatment systems, and nutrient management. The EMS is certified by a qualified third party.

MANURE MANAGEMENT ON U.S. HOG FARMS

One consequence of structural change in U.S. hog production has been the manure management challenge posed by concentrating more animals on a limited land base. Hog manure is primarily handled in two types of storage structures, lagoons and pits or tanks (see box, "Manure Storage and Handling Strategies"). Lagoons are large earthen containment structures into which manure and wastewater are flushed and maintained in liquid form until removed. Manure pits are often located under hog production facilities where, in the typical system, manure drops into pits through slatted floors and is stored in a slurry form until removed. These storage structures hold the manure until it can be land-applied on the same farm or nearby farms to meet crop nutrient needs. Technologies for land application include liquid/slurry manure spreaders that may or may not incorporate manure into the soil at application and irrigation systems that spread the liquid lagoon solution on nearby fields.

The different systems for manure management have very different impacts on the nutrient content of the manure, primarily nitrogen, and thus on the amount of land needed to spread manure (McBride and Key, 2003). For example, handling manure in pit or tank storage and using slurry spreaders to inject the manure into the soil manages the manure for its potential fertilizer value. This system is designed to retain manure nitrogen for crop use, and thus it requires more land on which to apply the manure if the operation is following a nitrogen-based nutrient management plan. In contrast, handling manure from lagoon storage and distributing it with irrigation increases the release of nitrogen into the atmosphere, reducing the manure's nutrient content and requiring less land for application.

Manure Storage and Handling

Lagoon use has a strong positive association with the scale of production (figure 5). For example, in 2009, 41 percent of operations with at least 1,000 head used a lagoon, compared with only 22 percent of small-scale operations and 27 percent of medium-scale operations. Despite the trend toward larger operations, there was a shift between 1998 and 2009 toward the use of pit/tank systems. By 2009, 62 percent of hogs were raised on farms using pit/tanks (up from 37 percent in 1998); in 2009, 34 percent were raised on farms using a lagoon system (down from 55 percent in 1998).[5] This shift can be partly attributed to regional shifts in hog production. In the Southern region, more hogs are produced on operations that use lagoon systems, while in other regions more hogs are produced on operations using pit/tanks (figure 3). Hence, the largely stagnant hog production in the South, combined with an increase of production in the Heartland between 1998 and 2009 (figure 4), partly explains the increasing national use of pit/tanks. The shift to pit systems may also have been encouraged by rising prices of chemical fertilizers, which increased incentives for new hog facilities to adopt designs that conserve manure nutrients for use on nearby crop or hay land.

The manure storage facility generally determines how manure effluent is transported and applied to crops. Pit/tank systems commonly use a slurry or liquid spreader, while irrigation technology (solid-set sprinklers, traveling guns, and linear or circular traveling booms) is used to move and apply lagoon liquid. The method of applying manure can have important implications for air quality, affecting the level of odorous gases (ammonia and hydrogen sulfide), particulate matter (including byproducts of ammonia), and greenhouse gases (methane and nitrous oxide) (Aillery et al., 2005).

Both solid and liquid manure can be incorporated into the soil, which reduces odor and nutrient volatilization (escape into the atmosphere) relative to spreading, keeping more nutrients available for plant uptake. Incorporation also reduces the risk of nutrient runoff. Sprinkler application increases nitrogen volatilization, which reduces the nitrogen available for plant use. Lagoon/sprinkler systems allow producers to dispose of manure from a given operation on fewer acres when a nitrogen criterion is used to determine application levels.

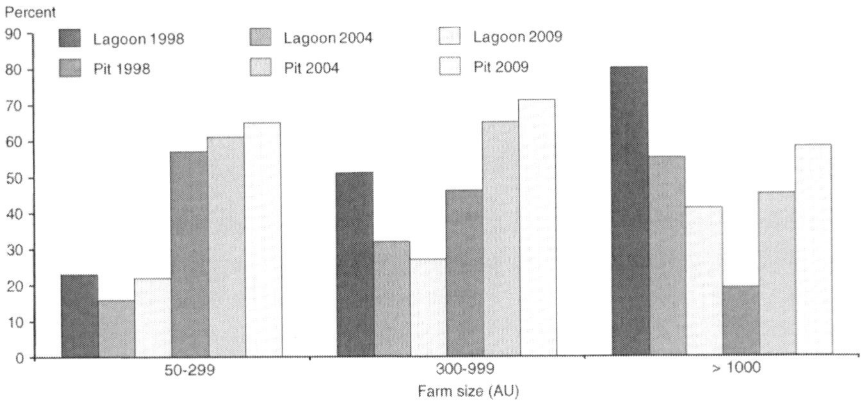

Note: Animal units (AU) are defined as 1,000 pounds of live animal weight, and the inventory of animal units is based on an estimate of the average number of hogs and pigs on the operation in each year.

Source: USDA, ERS, 1998, 2004, and 2009 Agricultural Resource Management Surveys.

Figure 5. Changes in the share of operations having lagoon and pit manure systems, 1998 to 2009.

MANURE STORAGE AND HANDLING STRATEGIES

Comprehensive nutrient management plan (CNMP)-Following a comprehensive nutrient management plan for applying manure and commercial fertilizer to land can reduce losses of nutrients to water resources through runoff or leaching (USDA, NRCS, 2005).

Nutrient management matches applications to crop needs so that as few nutrients as possible are lost to the environment. A CNMP is a group of conservation practices and management activities that ensure that both production and natural resource protection goals are achieved. Specific elements of a CNMP include background and site information; plans for manure and wastewater handling and storage, farm safety and security, land treatment practice, and prevention of soil erosion; nitrogen and phosphorus risk assessment; plans for nutrient management in accord with criteria in the Nutrient Management Conservation Practice (code 590); and record-keeping. Important objectives in comprehensive nutrient management planning include soil and manure testing for nutrient content, balancing farm-available nutrient resources with farm crop needs, and monitoring the operation's total nutrient balance to account for nutrients generated, field-applied, removed in products, or transferred offsite. Plans can also account for atmospheric losses of nitrogen, as well as for atmospheric deposition of nitrogen on cropland.

Manure incorporation and injection - Rapidly incorporating manure into the soil, either by plowing or disking solids after spreading or injecting liquids and slurries directly into the soil, reduces odor and gaseous emissions (Abt, 2000; Arogo et al., 2002). It also reduces the risk of nutrients being transferred to adjacent water bodies.

Slurry pits - Slurry systems store undiluted, untreated manure in water- tight tanks or pits until it can be land-applied. Storage can be either under the house or outdoors. The stored slurry is surface-applied to fields by sprayer trucks or wagons, incorporated into the soil with chisel plows behind nurse tanks, or directly injected into the soil with drag hoses (U.S. EPA, 2004a).

Lagoons - Lagoon systems use holding ponds to treat diluted manure for an extended period of time. Lagoons stabilize organic matter, reduce the nutrient mass that must be land-applied, and vent a large quantity of the manure nitrogen as ammonia. Some of the diluted lagoon liquid is used to flush the production houses. The "digested" lagoon liquid is eventually sprayed on cropland. Lagoons are used primarily in warmer climates where the anaerobic processes can take place year-round (U.S. EPA, 2004a).

Table 2. Hog manure application technologies used on farms applying manure

Application technology	1998	2004	2009
All farms that apply manure: Solid spreader	64	45**	34**
Liquid spreader (no injection)	27	18**	19**
Liquid spreader (injection)	20	21	25
Irrigation	12	13	19*
All farms that apply manure, weighted by animal units:			
Solid spreader	36	19**	14**
Liquid spreader (no injection)	25	17*	12**
Liquid spreader (injection)	30	34	34
Irrigation	34	36	35
Farm size category (farms that apply manure): Small (50 – 299 animal units)			
Solid spreader	66	40**	38**
Liquid spreader (no injection)	40	28**	38
Liquid spreader (injection)	28	31	23
Irrigation	9	10	16**
Medium (300 – 999 animal units):			
Solid spreader	32	23	15**
Liquid spreader (no injection)	28	19	14*
liquid spreader (injection)	42	37	42
Irrigation	32	28	21
Large (> 1,000 animal units):			
Solid spreader	10	10	10
Liquid spreader (no injection)	7	12	8
Liquid spreader (injection)	20	30	30
Irrigation	58	57	43

Note: Asterisks indicate level of significance for the test of the null hypothesis that mean equals 1998 mean: ** =5%, * = 10%. Some operations may have used more than one technology or none of the technologies. Therefore, the columns may add up to more than or less than 100 percent.

Source: USDA, ERS, 1998, 2004, and 2009 Agricultural Resource Management Surveys.

The type of manure-spreading technologies used is often related to the scale of production. Among large farms that applied manure to crops, irrigation was the most common form of manure application, followed by injection of liquid manure (table 2). In contrast, smaller operations were more likely to spread solid manure, or to spread liquid manure without injection. Between 1998 and 2009, the share of appliers who spread solid manure declined overall, from 64 percent of operations in 1998 to only 34 percent in 2009. Most of this change occurred because (1) the number of small farms, which are more likely to handle solid manure, declined and (2) small and medium farms handled manure in solid form less often.

Growers altered their spreading technologies between 1998 and 2009 toward methods that produce less odor, nutrient volatilization, and runoff. The share of animal units on operations where liquid manure was applied without soil injection declined from 25 to 12 percent. The share of large operations applying liquid manure with injection technologies increased 10 percentage points and the share of large operations applying manure with irrigation declined 15 percentage points (although neither of these are statistically significant).

Manure Application and Disposal

On average, larger hog operations harvested more cropland, applied manure to more acres on-farm, and applied manure to a larger share of their crop-land (table 3). These relationships are reflected in the aggregate statistics, as the average farm size increased between 1998 and 2009. Over this period, the average amount of cropland on a hog farm increased from 448 to 578 acres, and the number of manure-applied acres increased from 85 to 136. As a result, the share of cropland on hog farms on which manure was applied increased from 19.1 to 23.5 percent. This trend was mainly driven by the largest operations, where the share of cropland with manure application increased from 25 percent in 1998 to 30.9 percent in 2009.

Even though the share of on-farm cropland with manure application increased between 1998 and 2009, the average hog operation applied manure to less than a quarter of its cropland in 2009. This indicates that the potential exists to spread manure over more on-farm crop acreage. However, farmland is often separated by considerable distances from hog facilities, implying that operators could face substantially higher manure transportation and application costs if they were to apply manure on more acreage.

Table 3. Hog manure application technologies used on farms applying manure

	1998	2004	2009
All farms that apply manure:			
Acres with manure application	85	86	136**
Acres of cropland	448	483	578**
Percent of cropland with manure application	19.1	17.8	23.5*
Application intensity (AU/acre applied)	2.1	3.0**	3.7**
All farms that apply manure, weighted by animal units:			
Acres with manure application	147	218**	278**
Acres of cropland	596	855**	919**
Percent of cropland with manure application	24.7	25.5	30.2
Application intensity (AU/acre applied)	7.2	7.4	6.4
Farm size category (farms that apply manure) Small (50 – 299 animal units):			
Acres with manure application	95	85	101
Acres of cropland	517	599	592
Percent of cropland with manure application	18.4	14.2*	17.1
Application intensity (AU/acre applied)	1.3	1.6	1.5
Medium (300 – 999 animal units):			
Acres with manure application	156	169	178
Acres of cropland	565	652	661
Percent of cropland with manure application	27.6	26.0	27.0
Application intensity (AU/acre applied)	2.9	3.0	3.2
Large (> 1,000 animal units):			
Acres with manure application	159	224	281**
Acres of cropland	643	1016*	909
Percent of cropland with manure application	25.0	22.0	30.9
Application intensity (AU/acre applied)	9.4	8.0	6.2**

Note: Asterisks indicate level of significance for the test of the null hypothesis that mean equals 1998 mean: ** =5%, * = 10%.

Source: USDA, ERS, 1998, 2004 and 2009 Agricultural Resource Management Surveys.

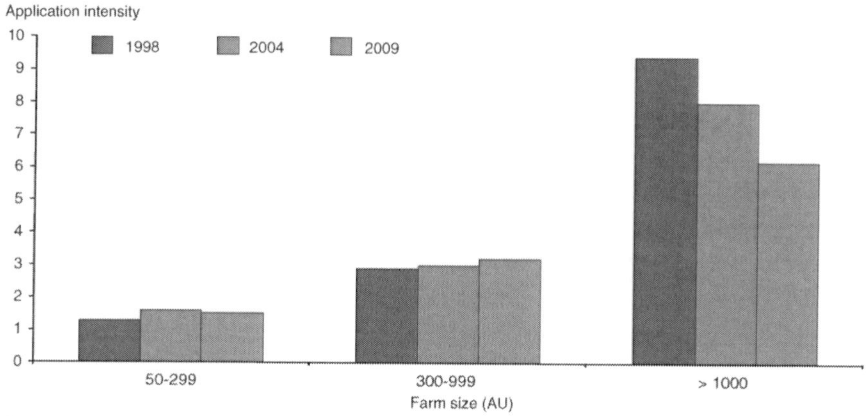

Note: Animal units (AU) are defined as 1,000 pounds of live animal weight, and the inventory of animal units is based on an estimate of the average number of hogs and pigs on the operation in each year. The manure application intensity is the hog inventory (AU) adjusted for the removal of manure off the farm divided by the acres of land on the hog operation on which manure was applied. For farms that moved manure off the operation, the number of AU was reduced by the equivalent amount of manure removed. For example, if 50 percent of the manure was moved off a 1,000-AU operation, only 500 AU was used to compute the ratio.

Source: USDA, ERS, 1998, 2004, and 2009 Agricultural Resource Management Surveys.

Figure 6. Manure application intensity increased with scale of production, but large-scale operations reduced intensity between 1998 and 2009.

Figure 6 illustrates the strong positive association between scale of production and manure application intensity (animal units per manure-applied acre, adjusted for the share of manure removed from the operations).6 The higher application rates for larger operations reflect the large amount of manure generated by larger hog operations compared with the cropland on these operations available for manure application. Between 1998 and 2009, the increase in total animal units produced outpaced the increase in crop acreage on which manure was applied, resulting in a 74-percent increase in average manure application intensity. However, this increase was driven mainly by operations with fewer than 1,000 animal units. For operations with more than 1,000 animal units—which are more likely to be subject to nutrient management restrictions—the application intensity actually declined by 34 percent (figure 6). Because large farms are responsible for a large share of hog output, the application intensity for the average unit of production (rather than the average farm) fell by 11 percent between 1998 and 2009.

Table 4. Manure application and removal

	Percent		
	1998	**2004**	**2009**
All farms:			
Applied manure on farm	83	82	76*
Removed manure from operation	14	21**	21**
Sold manure	0	2*	5**
Paid for manure removal	2	2	2
Manure given away free	12	18*	16*
All farms, weighted by animal units:			
Applied manure on farm	85	76	74*
Removed manure from operation	23	31	34**
Sold manure	1	5	9**
Paid for manure removal	4	4	5
Manure given away free	19	23	24
Farm size category Small (50 – 299 animal units):			
Applied manure on farm	89	89	83
Removed manure from operation	16	14	17
Sold manure	0	2	2**
Paid for manure removal	2	2	3
Manure given away free	14	11	13
Medium (300 – 999 animal units):			
Applied manure on farm	90	83	79**
Removed manure from operation	31	27	29
Sold manure	2	5	9
Paid for manure removal	8	5	3
Manure given away free	23	18	22
Large (> 1,000 animal units):			
Applied manure on farm	85	69	75
Removed manure from operation	26	38	37
Sold manure	1	3	9**
Paid for manure removal	1	3	1
Manure given away free	24	31	26

Note: Asterisks indicate level of significance for the test of the null hypothesis that mean equals 1998 mean: ** =5%, * = 10%.

Source: USDA, ERS, 1998, 2004 and 2009 Agricultural Resource Management Surveys.

Qualifying the measure of manure application intensity is important. Different storage and handling techniques help determine the quantity of nutrients in applied manure, so application intensity does not measure actual nutrient application rates. In addition, increases in feed efficiency have likely reduced the quantity of nutrients excreted by hogs. Nitrogen and phosphorus enter the production system in animal feed. Some of the nutrients are retained in the animal product (meat), but most are excreted in urine and manure (Follett and Hatfield, 2001; ASAE, 2005). In addition, feed efficiency is positively correlated with the scale of production – larger operations generally use less feed per hog produced. Hence, the nutrient application intensity (e.g., nitrogen per acre) is generally lower on larger farms than the estimated manure application intensity (animals per acre) would imply.

Among all hog operations, 76 percent applied some manure on their farms in 2009, a decrease from 82 percent in 2004 (table 4). The share of farms applying manure declined over this period for all size categories, particularly the medium- and large-size groups. Part of the overall decline can be explained by the shift to larger operations, which were less likely to apply manure on-farm than smaller operations in each of the three surveys. In 2009, 75 percent of large farms applied manure on-farm compared with 79 percent of medium-size farms and 83 percent of small farms.

Manure that is not applied on-farm is usually hauled off-farm for application to cropland on neighboring farms. Because large hog operations have relatively less cropland available on-farm, the relationship between the scale of production and the quantity of manure removed is positive (table 4). The overall share of farms removing manure grew 50 percent between 1998 and 2004, and this increase was attributable mainly to large operations. Between 2004 and 2009, the share of farms removing manure did not increase markedly. Most of the manure removed from farms was given away to nearby farms. However, there appears to be a growing trend toward selling manure—the share of farms selling manure increased in all farm size categories between 1998 and 2009, albeit from a very low beginning level. Larger operations were more likely to sell manure. In 2009, 9 percent of the largest operations sold manure, compared with only 1 percent in 1998. The increasing manure sales could be partly explained by the dramatic rise in chemical fertilizer prices, which increased 174 percent between 1998 and 2009.[7]

Manure Nutrient Management Practices

Table 5 describes the evolution of manure management practices between 1998 and 2009. Manure nutrient testing, a practice required as part of many State-mandated manure management plans, was positively associated with scale of production (figure 7). Larger operations are more likely to face State regulations that require nutrient management plans. The share of farms testing for nitrogen (N) increased from 18 to 29 percent between 1998 and 2004 and to 49 percent in 2009. The share of animal units on farms that tested manure for N increased from 51 percent to 73 percent to 86 percent over the three surveys. Nitrogen testing rates increased for all farm size categories, especially medium-scale operations (figure 7). Large operations did not have as much scope to increase their testing rate because 81 percent of these farms tested in 1998 (figure 7). The patterns for phosphorus (P) testing are similar to those displayed for N testing.

Table 5. Nutrient management practices

Management practices	Percent		
	1998	2004	2009
All farms:			
Tested manure for N content	18	29**	49**
Tested manure for P content	17	28**	48**
Applied commercial fertilizer and manure	61	58	50**
Applied manure to Bermuda grass (appliers only)	n.a.	11	7
Followed CNMP1	n.a.	30	55
Added microbial phytase to feed	4	13**	23**
All farms, weighted by animal units:			
Tested manure for N content	51	73**	86**
Tested manure for P content	50	72**	84**
Apply commercial fertilizer and manure	48	39*	44
Applied manure to Bermuda grass (appliers only)			
Followed CNMP[1]	n.a.	62	82
Added microbial phytase to feed	12	30**	39**

N=Nitrogen; P=Phosphorous. Asterisks indicate level of significance for the test of the null hypothesis that mean equals 1998 mean: ** =5%, * = 10%. n.a. indicates data not available. [1]CNMP is a comprehensive nutrient management plan (see box, "Manure Storage and Handling Strategies," page 12).

Source: USDA, ERS, 1998, 2004 and 2009 Agricultural Resource Management Surveys.

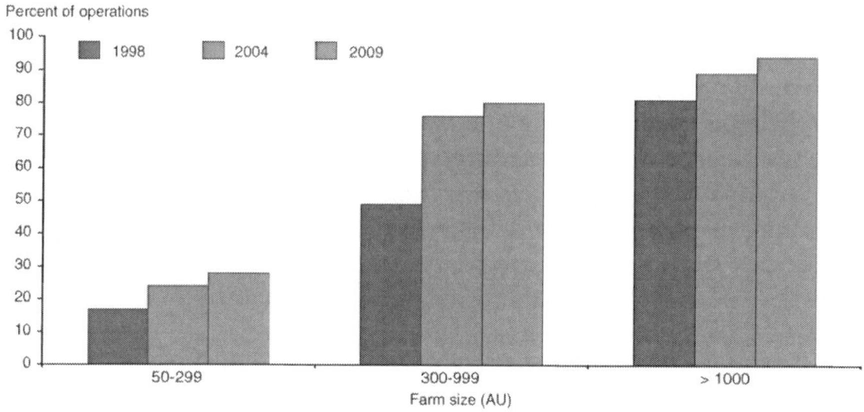

Note: Animal units (AU) are defined as 1,000 pounds of live animal weight, and the inventory of animal units is based on an estimate of the average number of hogs and pigs on the operation in each year.

Source: USDA, ERS, 1998, 2004, and 2009 Agricultural Resource Management Surveys.

Figure 7. Operations increased rate of manure nitrogen (N) testing between 1998 and 2009.

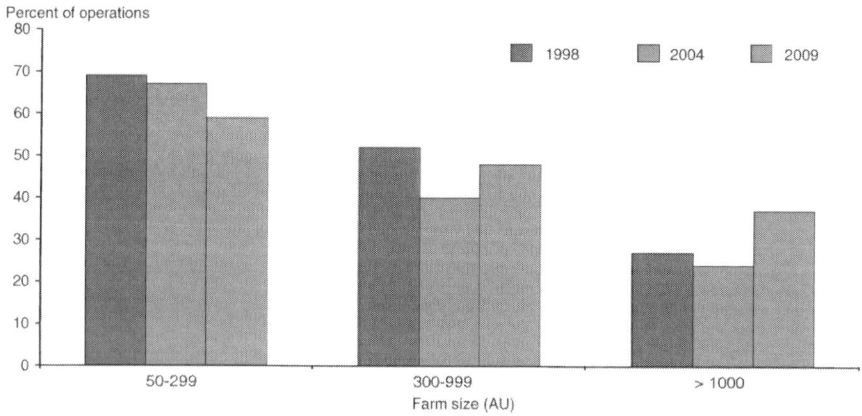

Note: Animal units (AU) are defined as 1,000 pounds of live animal weight, and the inventory of animal units is based on an estimate of the average number of hogs and pigs on the operation in each year.

Source: USDA, ERS, 1998, 2004, and 2009 Agricultural Resource Management Surveys.

Figure 8. Application of commercial fertilizer with manure declines with size.

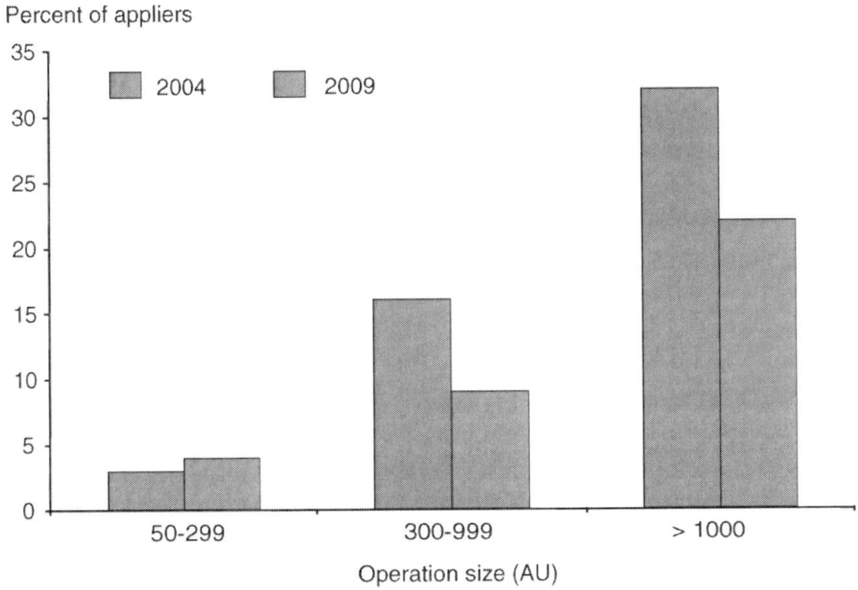

Note: Animal units (AU) are defined as 1,000 pounds of live animal weight, and the inventory of animal units is based on an estimate of the average number of hogs and pigs on the operation in each year.

Source: USDA, ERS, 1998, 2004, and 2009 Agricultural Resource Management Surveys.

Figure 9. Larger operations are more likely to apply manure to Bermuda grass 2004 and 2009.

Commercial fertilizer is applied to crops in addition to manure if the manure nutrients do not meet the nutritional needs of the crops. Testing the nutrient content of manure thus saves costs by avoiding the over application of supple- mental commercial fertilizer. Larger operations are more likely to have a surplus of nutrients provided by the manure produced on their operations and are therefore less likely to require supplemental commercial fertilizer. The association between the scale of production and the application of commercial fertilizer on acreage where manure was applied is negative (figure 8).

One strategy for increasing manure disposal on a limited land base is to plant crops that have a high rate of nutrient uptake. Bermuda grass, which is grown primarily in the South and Southeast, is especially appealing to hog producers because it consumes large amounts of nitrogen per acre. The scale of production and the application of manure to Bermuda grass had a strong positive association in 2004 and 2009 (figure 9). Between 2004 and 2009 (the only years for which the data are available), the share of operations applying manure to Bermuda grass dropped. This drop may be explained by the increasing use of manure as a substitute for commercial fertilizer—which increased in price by 106 percent in nominal terms over this period—and by the shift in the share of total production from the South and Southeast to the Heartland.

Microbial phytase is used as an additive in finishing hog diets to increase the absorption of organic phosphorus, meaning that supplemental inorganic or mineral phosphorus may not be needed and feed costs are reduced. Increases in the price of phosphorus over the study period boosted incentives to add phytase to feed to reduce phosphorus supplements. In addition, phytase use reduces phosphorus excretion in manure. When application rates are limited by phosphorus as opposed to nitrogen, lower manure phosphorus content can lead to increased spreading options by reducing the acres required to safely absorb manure nutrients. Farms with at least 300 animal units were much more likely to use microbial phytase than were smaller scale operations (figure 10). The share of farms using phytase grew in all size categories between 1998 and 2009, with the overall share of farms using phytase increasing from 4 to 23 percent. The share of hogs raised on farms using phytase increased from 12 to 39 percent. Concern about manure phosphorus levels is also evident from the increase in the share of farms testing manure for phosphorus content between 1998 and 2009 (table 5).

In 2004, about 30 percent of all farms followed a CNMP that required growers to apply manure nitrogen at or below the agronomic rate (see box, "Manure Storage and Handling Strategies").[8] By 2009, the share of farms following a CNMP had increased to 55 percent. The share of animal units raised on farms using a CNMP increased between 2004 and 2009 from 62 to 82 percent. The association between the scale of production and the use of a CNMP is positive (figure 11). About 46 percent of operations with 50-299 animal units followed a CNMP in 2009, compared with more than 80 percent of those with at least 300 animal units.

Manure Management for Air Quality

Air quality concerns related to manure include odorous gases (including ammonia and hydrogen sulfide), particulate material (byproducts of ammonia), and greenhouse gases (methane and nitrous oxide). Air pollutants are emitted from livestock buildings and manure management and treatment facilities, and from the land application of manure (National Research Council, 2003). Hydrogen sulfide and ammonia have been implicated in human and swine health problems (Carson, Osweiler, and Thorne, 2002; Donham, 2002). Modeled predictions suggest that the livestock sector is the source of a significant share of total U.S. ammonia emissions (EPA, 2000b). Individual health incidents involving livestock operation employees and public health findings pertaining to residents in the vicinity of livestock facilities provide indications of hog production's role in air pollution (Sneeringer, 2009; Cole, Todd, and Wing, 2000).

One of the challenges in addressing the environmental problems associated with livestock waste is the tradeoff between air and water pollution. Regulations designed to improve water quality may encourage manure management practices that result in greater air emissions, or vice versa (Aillery et al., 2005). For example, in order to reduce manure's nutrient content and meet regulatory nutrient application limits on cropland, operators may allow manure to decompose in uncovered lagoons rather than enclosed pits, which may result in more air pollution.

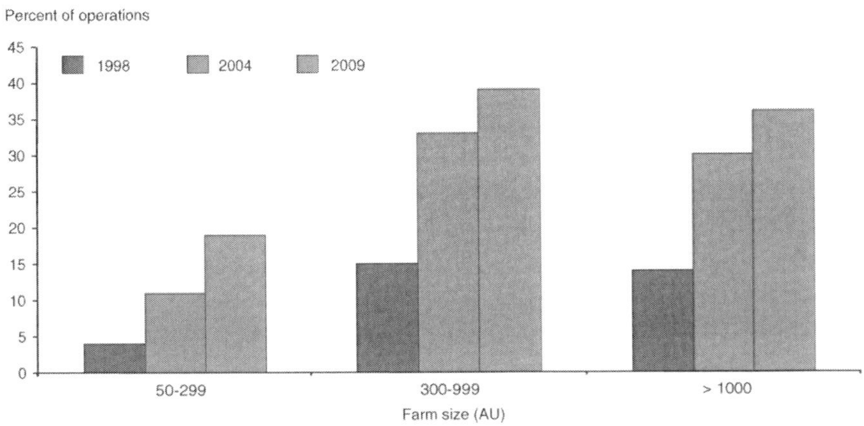

Note: Animal units (AU) are defined as 1,000 pounds of live animal weight, and the inventory of animal units is based on an estimate of the average number of hogs and pigs on the operation in each year.

Source: USDA, ERS, 1998, 2004, and 2009 Agricultural Resource Management Surveys.

Figure 10. Use of microbial phytase in feed increased between 1998 and 2009.

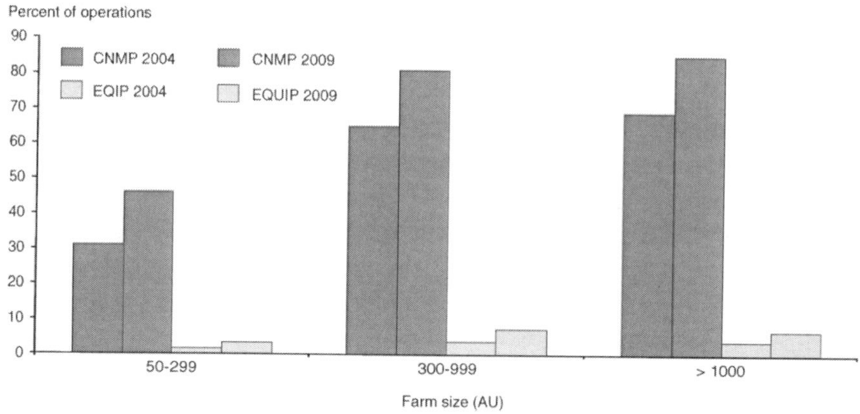

Note: Animal units (AU) are defined as 1,000 pounds of live animal weight, and the inventory of animal units is based on an estimate of the average number of hogs and pigs on the operation in each year.

Source: USDA, ERS, 1998, 2004, and 2009 Agricultural Resource Management Surveys.

Figure 11. More operations followed a CNMP and received EQIP payments in 2009.

The type of manure storage facility used affects the level of ammonia and other odorous gases. Lagoons reduce manure's nitrogen content through anaerobic digestion and ammonia volatilization. This allows farmers to apply more manure on less land without exceeding crop nutrient requirements, thereby lowering manure transportation costs by eliminating the need to transport manure to more distant cropland. In contrast, pit manure facilities, which conserve manure nutrients for use on cropland, emit less ammonia. It has been estimated that operations using lagoons emit almost twice as much ammonia per animal as operations using pit systems (Key and Kaplan, 2007).

Lagoon covers – plastic covers that float on the lagoon surface or that are tented over lagoons – can greatly decrease gaseous emissions, including ammonia (Jacobson et. al., 1999; Arogo et al., 2002). In 2009, 5 percent of the hog farms in the survey that had a lagoon used a cover to control odor and air emissions. Among hogs produced on operations with lagoons, only 4 percent were produced on operations with lagoon covers, suggesting that operations with covers are somewhat smaller than average. Covering manure storage tanks can also greatly reduce the discharge of ammonia to the atmosphere. Storage tanks can be covered with a roof, concrete lid, or flexible plastic cover. The surface of the stored waste can also be covered with straw or other materials (polystyrene foam, air-filled clay balls) that act as biofilters. While reducing ammonia emissions from the storage facility, lagoon and pit covers increase the nitrogen content of effluent that is eventually spread on fields. This raises the potential for both ammonia emissions and nitrate pollution of water resources from the manure applied to fields.

Some covered systems (anaerobic digesters) also capture methane and burn it as a biofuel to generate electricity. Methane is produced when livestock manure decomposes anaerobically (without oxygen). Lagoon and pit manure handling systems that are common on dairy and hog operations emit large amounts of methane. Methane is a potent greenhouse gas that has about 25 times the global warming potential of carbon dioxide. Burning the methane captured from an anaerobic digester greatly reduces the gas's global warming potential. (See box, "Anaerobic Digesters and Greenhouse Gas Emissions.)

Manure can be managed in a variety of ways to control odor. Chemicals can be added to manure during its collection in order to bind odorous compounds and reduce ammonia emissions. Trapping air vented from production houses and treating it before discharge to the atmosphere can also reduce odorous compounds. Table 6 displays the percentage of farms in 2004 and 2009 that used additives or filters to control odor. Between 2004 and 2009, the share of farms that added chemicals to manure rose from 13 to 17 percent, while the share of animals on operations that added chemicals fell slightly, from 31 to 29 percent. About 5 percent of operations used filters to ventilate barns to control odor in both years, although the share of livestock produced by these operations decreased slightly, from 10 to 8 percent.

Environmental Policy Effects on Manure Management

Recent policy initiatives may explain some of the changes in manure management practices. Federal and State policies implemented in recent years have set limits on the amount of nutrients that can be applied per acre of land. Restricting application rates may help explain increases in the crop acreage receiving animal manure and the amount of manure moved off the farm, as well as the widespread adoption of nutrient management plans observed in the ARMS.

States have adopted the updated 2008 Federal NPDES rules regarding CAFOs at different paces, creating variation in regulation status. Consequently, examining how manure management practices vary according to a State's degree of regulation adoption is possible. States are placed in one of three categories according to EPA's summary of the status of CAFO rule adoption (EPA, 2010b). The first category includes States that needed no further updates as of 2009 (this includes States that have not received NPDES program authorization from the EPA, meaning that the Federal regulations apply). Of the 19 States included in the ARMS survey, 3 fall in this category (Michigan, Minnesota, and Oklahoma). The majority of States in ARMS fall in the second category, which includes those completing their regulation updates in 2010. The third category consists of States that the EPA has listed with a regulatory update completion date "to be determined."

ANAEROBIC DIGESTERS AND GREENHOUSE GAS EMISSIONS

Anaerobic digesters, also known as "methane digesters," "biodigesters," or "biogas recovery systems, " can be used with lagoon or pit-type manure storage facilities to collect manure, optimize it for the production of methane by adjusting temperature and water content, capture the biogas, and burn it for heat or electricity generation. Methane digesters can provide numerous benefits to livestock producers and the environment. Digesters can supply a renewable source of electricity to power farm equipment or be sold to the electricity distribution grid. Digesters can reduce greenhouse gas emissions, odors from manure, and the potential for surface water contamination. They can also be used to recycle manure solids for animal bedding material. Currently, 157 methane digesters operate in the United States, of which 126 are on dairy operations and 24 are on hog operations (U.S. EPA, 2010a).

Digesters have received attention lately because of their ability to reduce greenhouse gas (GHG) emissions from livestock manure. Burning 1 ton of methane is equivalent to eliminating about 24 tons of carbon dioxide. If the emissions reductions from methane digesters meet certain criteria, farmers may be able to sell these GHG reductions, or "carbon offsets," to other GHG emitters who face emissions caps or who voluntarily wish to offset their own emissions. Currently, only a few U.S. livestock operators sell offsets in regional or voluntary carbon offset markets, partly because the price of carbon offsets in these markets has been low.

However, future efforts to reduce greenhouse gas emissions could result in substantially higher carbon prices, which could provide a new source of income for farmers who adopt methane digesters.

A recent ERS study found that with a moderate carbon price, revenues from the sale of carbon offsets could have a large effect on the number of hog operations that would adopt a methane digester. (See Key and Sneeringer, *Climate Change Policy and the Adoption of Methane Digesters on Livestock Operations*, Economic Research Report No. 111, http://www.ers.usda.gov/Publications/ERR111/ERR111.pdf for more information.)

Table 7 shows the percentage of farms following certain nutrient management practices by regulation adoption status. Farms in States with fully updated regulations appear to be the most active in pursuing nutrient management practices, followed by States with planned completion in 2010. For example, 62 percent of operations in States with fully updated regulations follow a CNMP, compared with 55 percent in States that planned to complete updating in 2010 and only 38 percent in States that had not updated by 2010.

Financial assistance from USDA's Environmental Quality Incentives Program (EQIP) can help defray the costs of meeting environmental regulations by funding planning, installation, maintenance, and technical support for protective conservation practices. Between 2004 and 2009, the share of farms in each size category receiving some EQIP payments related to hog production increased (figure 11). The overall average share of farms receiving EQIP payments increased from 1.5 to 4.4 percent between 2004 and 2009, and the share of animal units raised on farms receiving EQIP payments increased from 3.2 to 7.1 percent (table 8). EQIP payments were used primarily for installing manure handling and storage facilities and for developing and maintaining a nutrient management plan. While the number of operations that received payments was relatively small, these payments may have facilitated some of the changes observed in the study, such as the increased use of liquid spreading with injection, growth in the amount of cropland on which manure is applied, greater use of manure nutrient testing, greater use of microbial phytase in feed, and more widespread use of CNMPs.

Policy initiatives may partially explain the increased use of such practices as manure injection and development of a nutrient management plan, but agricultural-residential conflicts at the rural-urban fringe may also play a role (Ribaudo and Johansson, 2007). Manure injection reduces odors from land application, and developing a nutrient management plan demonstrates due care on the part of livestock producers in avoiding harm to the nearby community.

Table 6. Practices used to control odor, 2004 and 2009

Management practices	Percent	
	2004	2009
All farms:		
Manure additives	13	17
Filters for barn ventilation	5	5
All farms, weighted by animal units:		
Manure additives	31	29
Filters for barn ventilation	10	8

Note: Asterisks indicate level of significance for the test of the null hypothesis that mean equals 2004 mean: ** =5%, * = 10%.

Source: USDA, ERS, 2004 and 2009 Agricultural Resource Management Surveys.

Table 7. Manure management methods by EPA State regulatory status, 2009

		(a)	(b)	(c)
	All States with data in ARMS	Fully updated regulations	Planned completion of update in 2010	Completion of update of regulations to be determined
Number of States	19	3	11	5
Number of farms	24,421	3,998	18,003	2,419
Number of animal units (1,000)	13,313	2,425	10,318	571
	Percent			
All farms:				
Followed CNMP[1]	55	62[c]	55[c]	38[ab]
Test manure for N content	37	43	38	24
Test manure for P content	37	43	37	24
Adjust manure application according to N content	29	34	28	25
Adjust manure application according to P content	26	30	26	24

		(a)	(b)	(c)
	All States with data in ARMS	**Fully updated regulations**	**Planned completion of update in 2010**	**Completion of update of regulations to be determined**
All farms, weighted by animal units:				
Follow CNMP	82	88	80	78
Test for N content	64	69	62	69
Test for P content	63	69	61	69
Adjust manure application according to N content	31	46[b]	28[a]	24
Adjust manure application according to P content	28	41	25	24

[1] CNMP = Comprehensive Nutrient Management Plan (see box, "Manure Storage and Handling Strategies").

[a] Denotes statistically different at the 5% level from column (a).

[b] Denotes statistically different at the 5% level from column (b).

[c] Denotes statistically different at the 5% level from column (c).

Note: States with "fully updated regulations" include MI, MN, and OK (OK does not have NPDES program authorization and thus is directly regulated by the Environmental Protection Agency; it is therefore assumed to be up to date with EPA regulations); States with "planned completion of update in 2010" include PA, GA, KY, NC, IL, IN, OH, IA, MO, NE, and CO; States with "completion of update of regulations to be determined" include VA, WI, AR, KS, SD.

Source: USDA, ERS 2009 Agricultural Resource Management Survey.

CONCLUSION

Manure management practices and outcomes have changed significantly over the past decade. Many of these trends can be attributed to the pronounced structural changes and regional shifts in hog production that occurred between 1998 and 2009. In particular, the relatively strong production growth in the Heartland compared with that of the South likely explains much of the shift from lagoons to pit/tank systems, despite lagoons being more prevalent among larger operations. Another important structural change has been growth in farm size, which has contributed to the decline in the use of solid manure spreading.

Table 8. Environmental Quality Incentives Program (EQIP) payments related to hog production, 2004

	Percent	
	2004	2009
All farms:		
Any hog-related EQIP payments	1.5	4.4**
Manure handling and storage facilities	0.6	1.9*
Nutrient management plan	0.8	2.1*
Manure application	0.2	1.3**
Other[1]	0.4	2.0**
All farms, weighted by animal units:		
Any hog-related EQIP payments	3.2	7.1**
Manure handling and storage facilities	1.5	2.8
Nutrient management plan	2.2	4.0
Manure application	0.6	1.8
Other[1]	1.1	3.3*

[1] Includes animal facilities, manure hauling, and unspecified.

Note: Asterisks indicate level of significance for the test of the null hypothesis that mean equals 2004 mean: ** =5%, * = 10%.

Source: USDA, ERS, 2004 and 2009 Agricultural Resource Management Surveys.

The shift to larger operations has meant that an increasing share of production falls under the purview of regulations governing the application of manure nutrients to cropland. Larger farms with less cropland available per head for spreading manure are more likely to remove manure from the operation; are less likely to apply both commercial fertilizer and manure to crops and more likely to apply manure to crops with a high rate of nutrient uptake (e.g., Bermuda grass); and more likely to add microbial phytase to feed, to test soil for nutrients, and to follow a comprehensive nutrient management plan.

While the manure-nutrient application intensity generally increases with farm size, the decline in application intensity on the largest operations between 1998 and 2009 suggests that environmental policy is contributing to the adoption of conservation-compatible manure management practices. The number of Federal and State policies designed to reduce over application of manure nutrients grew between 1998 and 2009, and the number of States implementing and enforcing Federal policies increased. In 2009, 55 percent of farms, representing 82 percent of animal units, followed a nutrient management plan – a substantial increase from 2004, when only 30 percent of farms followed a CNMP. The share of farms receiving EQIP payments related to manure handling, storage, and application also increased substantially over this period.

Structural changes in hog production, combined with increasing fertilizer prices, an increase in regulations, stricter enforcement of regulations, and more generous cost-share programs, jointly explain many of the changes in hog manure management between 1998 and 2009, including:

- An increase in the share of hog operation cropland receiving manure;
- An increase in total cropland receiving manure per animal unit on hog operations;
- A decline in spreading solid manure and in spreading liquid manure without injection, among hog farms applying manure;
- An increase in the share of farms removing manure from their operations;
- An increase in manure nutrient testing rates; and
- Increased use of microbial phytase in feed.

Since 2004, higher chemical fertilizer prices have raised the value of manure as a source of crop nutrients. In response, producers applied more manure to crops on-farm and applied less to Bermuda grass. In addition, a greater share of producers sold manure.

The increasing concentration of hog production on large operations is expected to continue, meaning that manure management will remain an important issue for the hog industry and others concerned with its environmental impacts. The findings show that there is still significant room for reducing the impact of manure disposal practices; for example, hog operations, on average, apply manure to only about 30 percent of their available crop acreage. The research suggests that in addition to the influence of market and structural forces, hog producers have responded to policy incentives that address environmental side effects of manure management. Policy incentives, along with technological innovations, are likely to play an important role in the future of hog manure management and its environmental impacts.

References

Abt Associates. *Air Quality Impacts of Livestock Waste.* Report prepared for U.S. EPA Office of Policy, Bethesda, MD. 2000.

Aillery, M; Gollehon, N; Johansson, R; Kaplan, J; Key, N; Ribaudo, M. *Managing Manure To Improve Air and Water Quality.* U.S. Department of Agriculture, Economic Research Service. ERR-9, Sept. 2005.

Arogo, J; Westerman, PW; Heber, AJ; Robarge, WP; Classen, JJ. *Ammonia Emissions from Animal Feeding Operations.* National Center for Manure and Animal Waste Management, North Carolina State University, Raleigh. 2002.

ASAE. *Manure Production and Characteristics.* American Society of Agricultural Engineers, ASAE D384.2MAR2005. 2005.

Bales, JD; Oblinger, CJ; Sallenger, AH. Jr. *Two Months of Flooding in Eastern North Carolina, September-October 1999: Hydrologic, Water Quality, and Geologic Effects of Hurricanes Dennis, Floyd, and Irene.* U.S. Geological Survey. WRIR 00-4093, Raleigh, NC. 2000.

Bergstrom, JC; Centner, TJ. "Agricultural Nuisances and Right-To-Farm Laws: Implications of Changing Liability Rules," *Review of Regional Studies*, 19(Summer 1989), 23-30.

Carson, TL; Osweiler, GD; Thorne, PS. "Adverse Health Effects: Toxicology" in *Iowa Concentrated Animal Feeding Operations Air Quality Study.* Iowa State University and The University of Iowa Study Group. 2002.

Centner, TJ. "Agricultural Nuisances: Qualifying Legislative 'Right-to- Farm' Protection Through Qualifying Management Practices," *Land Use Policy*, 19 (July 2002), 259-67.

Cole D; Todd, L; Wing, S. "Concentrated Swine Feeding Operations and Public Health: A Review of Occupational and Community Health Effects," *Environmental Health Perspectives*, 108, 686-99. 2000.

Copeland, C. *Animal Waste and Hazardous Substances: Current Laws and Legislative Issues.* Congressional Research Service, Jan. 2008.

Donham, K. "The Concentration of Swine Production: Effects on Swine Health, Productivity, Human Health, and the Environment," *Toxicology*, 16, 559-97. 2000.

Duke, JM; Malcolm, SA. "Legal Risk in Agriculture: Right-to-Farm Laws and Institutional Change," *Agricultural Systems*, 75 (Feb.-Mar. 2003), 295-303.

Follett, RF; Hatfield, JL. *Nitrogen in the Environment: Sources, Problems, and Management*, Elsevier Press, New York, NY. 2001.

Jacobson, LD; Kozier, JA; Hoff, SJ; Heber, AJ; Parker, DB. "*Odor Emissions and Chemical Analysis of Odorous Compounds from Animal Buildings*." Paper presented at Workshop on Agricultural Air Quality: State of the Science, Potomac, MD, June 5-8, 2006.

Key, N; Kaplan, J. "Multiple Environmental Externalities and Manure Management Policy," *Journal of Agricultural and Resource Economics*, 32(1), (April 2007),115-34.

Key, N; McBride, W. *The Changing Economics of U.S. Hog Production*. U.S. Department of Agriculture, Economic Research Service, ERR-52, Dec. 2007.

Key, N; McBride, WD; Mosheim, R. "Decomposition of Total Factor Productivity Change in the U.S. Hog Industry." *Journal of Agricultural and Applied Economics*, 40(1), (April 2008), 137-49.

Key, N; Sneeringer, S. *Climate Change Policy and the Adoption of Methane Digesters on Livestock Operations*. U.S. Department of Agriculture, Economic Research Service, ERR-111. Feb. 2011. Accessed at: http://www.ers.usda.gov/Publications/ERR111/.

McBride, W; Key, N. *Economic and Structural Relationships in U.S. Hog Production*. U.S. Department of Agriculture, Economic Research Service, AER-818, Feb. 2003.

National Research Council, Ad Hoc Committee on Air Emissions from Animal Feeding Operations, Committee on Animal Nutrition. *Air Emissions from Animal Feeding Operations: Current Knowledge, Future Needs*. Washington, DC: National Academies Press, 2003.

North Carolina General Assembly. The Clean Water Responsibility and Environmentally Sound Policy Act, 1997. http://www.ncga.state gascripts/BillLookUp/BillLookUp.pl?Session=1997&BillID=H515,

Ogishi, A; Zilberman, D; Metcalfe, M. "Integrated Agribusinesses and Liability for Animal Waste," *Environmental Science and Policy*, 6(2), (2003),181-88.

Patton, M; Seidl, A. *Matrix of State Level Policies for Animal Feeding Operations*. Agricultural and Resource Policy Report APR99-01, Dept. of Agricultural and Resource Economics, Colorado State University, Ft. Collins, CO, Jan. 1999.

Redwine, JS; Lacey, RE. "A Summary of State-by-State Regulation of Livestock Odor." Paper presented at Second International Conference on *Air Pollution from Agricultural Operations of the ASAE*, Des Moines, IA. Oct. 10, 2000.

Ribaudo, M; Cattaneo, A; Agapoff, J. "Cost of Meeting Manure Nutrient Application Standards in Hog Production: The Roles of EQIP and Fertilizer Offsets," *Review of Agricultural Economics*, 26(4), (2004), 430-44.

Ribaudo, MO; Johansson, R. "Nutrient Management Use at the Rural- Urban Fringe: Does Demand for Environmental Quality Play a Role?" *Review of Agricultural Economics*, 29(4), (2007), 689-99.

Sneeringer, S. "Does Animal Feeding Operation Pollution Hurt Public Health? A National Longitudinal Study of Health Externalities Identified by Geographic Shifts in Livestock Production," *American Journal of Agricultural Economics*, 91(1), 124-37. 2009.

Sneeringer, Stacy E. "A National, Longitudinal Study of the Effects of Concentrated Hog Production on Air Pollution," *American Journal of Agricultural Economics*, 92(3), 821-35, 2010.

U.S. Department of Agriculture, Natural Resources Conservation Service. "Draft Comprehensive Nutrient Management Planning Technical Guidance," *National Planning Procedures Handbook*, Subpart B, Part 600.51. June 2005. http://www.nrcs.usda.gov/technical/afo/ cnmp_guide_ index.html/.

U.S. Environmental Protection Agency. *Air Pollution Regulatory Assessment for Animal Feeding Operations*. Office of Air Quality Planning and Standards, Research Triangle Park, NC. 2000a.

U.S. Environmental Protection Agency. *National Air Pollution Emissions Trends, 1900-1998*. Research Triangle Park, NC. 2000b.

U.S. Environmental Protection Agency. *State Compendium of Programs and Regulatory Activities Related to Animal Feeding Operations.* Office of Wastewater Management, Water Permits Divisions, Washington, DC. 2002.

U.S. Environmental Protection Agency. *State Compendium: Programs and Regulatory Activities Related to Animal Feeding Operations.* Tetra Tech, Fairfax, VA. May 2002.

U.S. Environmental Protection Agency. *National Emission Inventory—Ammonia Emissions from Animal Husbandry Operations.* Draft Report, Jan. 30. 2004a. http://www.epa.gov/ttn/chief/ap42/ch09/related/ nh3inventorydraft_jan2004.pdf.

U.S. Environmental Protection Agency. *Recommendations for Areas To Be Designated "Nonattainment" for the Fine Particle National Air Quality Standards.* Office of Air Quality Planning and Standards, Dec. 2004b.

U.S. Environmental Protection Agency. "EPA Takes Important Step in Controlling Air Pollution from Farm Country Animal Feeding Operations," News release, August 22, 2006. http://yosemite.epa.gov/opa/admpress.nsf/a8f952395381d3968525701 c005e65b5/a3b628e23af32f68852 571d200618474!OpenDocument/.

U.S. Environmental Protection Agency. CERCLA/EPCRA Administrative Reporting Exemption for Air Releases of Hazardous Substances from Animal Waste at Farms: Final Rule. Office of Solid Waste and Emergency Response, Washington, DC, February. 2009a.

U.S. Environmental Protection Agency. Guide for the Agriculture and Livestock Sectors: Proposed Rule: Mandatory Reporting of Greenhouse Gases. EPA-430-F-09-059, Was-**hington, DC, June. 2009b. Accessed at http://www.epa.gov/climatechange/emissions/ghgrulemaking.html. November 21, 2010.

U.S. Environmental Protection Agency. U.S. Anaerobic Digester Status Report. October 2010a. Accessed at: http://www.epa.gov/agstar/documents/digester_status_report2010.pdf/.

U.S. Environmental Protection Agency. *NPDES CAFO Rule Implementation Status – National Summary, First Quarter 2010.* March 31, http://cfpub.epa.gov/npdes/whatsnew.cfm?program_id=7/.

U.S. Environmental Protection Agency. "Emissions Data from Animal Feeding Operations Study Now Available / EPA also solicits additional information to further understand emissions," News release, Jan. 13, 2011. http://yosemite.epa.gov/opa/admpress.nsf/d0cf6618525a9efb8525 7359003fb69d/dc13b657ff6203ce85257817005ed001!OpenDocument/.

Wall, E; Weersink, A; Swanton, C. "Agriculture and ISO 14000," *Food Policy*, 26(1), 35-48, 2001.

Williams, M. "*Progress of Identifying and Installing Potential Environmentally Superior Technologies per Agreements between the North Carolina Attorney General, Smithfield Foods, Premium Standard Farms, and Frontline Farmers.*" Presentation at public meeting, North Carolina State University, Raleigh, NC, July 14, 2004.

End Notes

[1] Cropland includes land used for hay production, but excludes land used for pasture only.

[2] As farms have grown in size, some have divided into multiple operations, each specializing in hog or crop production. These hog operations selected in the ARMS would have reported little if any crop acreage because the crop acreage was part of the other operation.

Partnerships between family members may organize farms by specializing operations in this manner. This arrangement lowers the estimated crop acreage per farm and increases the percent of hog operations with no crops. The extent of this influence is unknown, but it should not affect the inter year comparison of hog farms.

[3] The EPA report does not specify which types of livestock facilities will be covered.

[4] For more information about EQIP see the USDA Natural Resource Conservation Service website: http:// www.nrcs.usda.gov/programs/eqip/.

[5] In tables 2–7, "all farms, weighted by animal units" gives the mean values computed using a weight defined as the sample weight times the animal units on the operation. This weighted mean describes the manure system used on the average animal unit rather than the average farm.

[6] For the intensity ratio, the denominator is the acres of land on the hog operation on which manure was applied. The numerator is the farm inventory (AU), adjusted for the removal of manure off the farm. For farms that moved manure off the operation, the number of AU was reduced by the equivalent amount of manure removed. For example, if 50 percent of the manure was moved off a 1,000-AU operation, only 500 AU was used to compute the ratio.

[7] From "Fertilizer Use and Price Tables: Fertilizer Price Indexes," USDA Economic Research Service. Accessed November 2010 at: www.ers.usda.gov/Data/FertilizerUse/ Tables/ Table8.xls.

[8] The goal of manure application is to apply manure at rates that meet crop needs while avoiding overapplication that could impair water quality. Rates that meet this goal are often called agronomic rates.

In: Manure Management
Editor: Sheila E. Corby
ISBN: 978-1-62948-874-5

Chapter 3

ANIMAL WASTE AND HAZARDOUS SUBSTANCES: CURRENT LAWS AND LEGISLATIVE ISSUES[*]

Claudia Copeland

SUMMARY

The animal sector of agriculture has undergone major changes in the last several decades: organizational changes within the industry to enhance economic efficiency have resulted in larger confined production facilities that often are geographically concentrated. These changes, in turn, have given rise to concerns over the management of animal wastes and potential impacts on environmental quality.

Federal environmental law does not regulate all agricultural activities, but certain large animal feeding operations (AFOs) where animals are housed and raised in confinement are subject to regulation. The issue of applicability of these laws to livestock and poultry operations— especially the Comprehensive Environmental Response, Compensation, and Liability Act (CERCLA, the

[*] This is an edited, reformatted and augmented version of a Congressional Research Service Publication, CRS Report for Congress RL33691, dated September 10, 2013.

Superfund law) and the Emergency Planning and Community Right-to-Know Act (EPCRA)—has been controversial and has drawn congressional attention.

Both Superfund and EPCRA have reporting requirements that are triggered when specified quantities of certain substances are released to the environment. In addition, Superfund authorizes federal cleanup of releases of hazardous substances, pollutants, or contaminants and imposes strict liability for cleanup and injuries to natural resources from releases of hazardous substances.

Superfund and EPCRA include citizen suit provisions that have been used to sue poultry producers and swine operations for violations of those laws. In two cases, environmental advocates claimed that AFO operators had failed to report ammonia emissions, in violation of Superfund and EPCRA. In both cases, federal courts supported broad interpretation of key terms defining applicability of the laws' reporting requirements. Three other cases not dealing with reporting violations also have attracted attention, in part because of questions of whether animal wastes contain hazardous substances that can create cleanup and natural resource damage liability under Superfund.

In December 2008, EPA issued a rule to exempt animal waste emissions to the air from most CERCLA and EPCRA reporting requirements. Legal challenges to the rule followed. In October 2010, a federal court approved the government's request to remand the rule to EPA for reconsideration and possible modification, but the agency has not yet proposed a new or revised rule.

The lawsuits testing the applicability of CERCLA and EPCRA to poultry and livestock operations and potential changes by EPA to the 2008 exemption rule have led to congressional interest in these issues. In the 112th Congress, legislation was introduced that would amend CERCLA to clarify that manure is not a hazardous substance, pollutant, or contaminant under that act and that the notification requirements of both laws would not apply to releases of manure (H.R. 2997 and S. 1729). Proponents have argued that Congress did not intend that either of these laws apply to agriculture and that enforcement and regulatory mechanisms under other laws are adequate to address environmental releases from animal agriculture. Opponents respond that enacting an exemption would severely hamper the ability of government and citizens to know about and respond to releases of hazardous substances caused by an animal agriculture operation. Similar legislation has not been introduced in the 113th Congress.

INTRODUCTION

The animal sector of agriculture has undergone major changes in the last several decades, a fact that has drawn the attention of policymakers and the public. In particular, organizational changes within the industry to enhance economic efficiency have resulted in larger confined production facilities that often are geographically concentrated.[1] Increased facility size and regional concentration of livestock and poultry operations have, in turn, given rise to concerns over the management of animal wastes from these facilities and potential impacts on environmental quality, public health and welfare.

Animal manure can be and frequently is used beneficially on farms to fertilize crops and add or restore nutrients to soil. However, animal waste, if not properly managed, can adversely impact water quality through surface runoff and erosion, direct discharges to surface waters, spills and other dry-weather discharges, and leaching into soil and ground. It can also result in emission to the air of particles and gases such as ammonia, hydrogen sulfide, and volatile organic chemicals. According to the U.S. Department of Agriculture (USDA), in 1997, 66,000 operations had farm- level excess nitrogen (an imbalance between the quantity of manure nutrients produced on the farm and assimilative capacity of the soil on that farm), and 89,000 had farm-level excess phosphorus.[2] USDA believes that where manure nutrients exceed the assimilative capacity of a region, the potential is high for runoff, leaching of nutrients, and other environmental problems. Geographically, areas with excess farm-level nutrients correspond to areas with increasing numbers of confined animals.

Federal environmental law does not regulate all agricultural activities. Some laws specifically exempt agriculture from regulatory provisions, and others are structured so that farms escape most, if not all, of the regulatory impact. Still, certain large animal feeding operations (AFOs) where animals are kept and raised in confinement are subject to environmental regulation. The primary regulatory focus on environmental impacts has been on protecting water resources and has occurred under the Clean Water Act. In addition, facilities that emit large quantities of air pollutants may be regulated under the Clean Air Act. Some livestock operations also may be subject to requirements of the Comprehensive Environmental Response, Compensation, and Liability Act (CERCLA, the Superfund law) and the Emergency Planning and Community Right- to-Know Act (EPCRA).[3] The issue of applicability of these laws to livestock and poultry operations—especially CERCLA and EPCRA—has been controversial and has drawn congressional attention.

This chapter describes the provisions of Superfund and EPCRA, and enforcement actions under these laws that have received attention. Congressional scrutiny in the form of legislative proposals and oversight hearings is discussed. Bills intended to exempt animal manure from the requirements of Superfund and EPCRA have been introduced several times since the 109th Congress, including in the 112th Congress (H.R. 2997 and S. 1729). Issues raised by the legislation are analyzed.

CERCLA AND EPCRA

Both the Comprehensive Environmental Response, Compensation, and Liability Act (the Superfund law, 42 U.S.C. §§9601-9675) and the Emergency Planning and Community Right-to- Know Act (42 U.S.C. §§11001-11050) have reporting requirements that are triggered when specified quantities of certain substances are released to the environment.[4] Both laws, which are administered by the Environmental Protection Agency (EPA), utilize information disclosure in order to increase the information available to government and citizens about the sources and magnitude of chemical releases to the environment. In addition to reporting requirements, CERCLA includes provisions authorizing federal cleanup of releases of hazardous substances, pollutants, or contaminants that may present an imminent and substantial danger to the public health or welfare (§104), and imposing strict liability for cleanup and damages for injury to, destruction of, or loss of natural resources resulting from releases of hazardous substances (§107). At issue today is how the reporting requirements and other provisions of these laws apply to poultry and livestock operations.

Superfund authorizes programs to remediate uncontrolled or abandoned hazardous waste sites and assigns liability for the associated costs of cleanup. Section 103(a) of CERCLA requires that the person in charge of a facility (as defined in §101(9)) that releases a "reportable quantity" of certain hazardous substances must provide notification of the release to the National Response Center.

EPCRA establishes requirements for emergency planning and notification for storage and release of hazardous and toxic chemicals. Section 304(a)(1) of EPCRA requires the owner or operator of a facility (as defined in §329(4)) to report to state and local authorities any releases greater than the reportable quantity of substances deemed hazardous under Superfund or extremely hazardous under EPCRA. Under Superfund, the term "release" (§101(22)) includes discharges of substances to water and land and emissions to the air from "spilling, leaking, pumping, pouring, emitting, emptying, discharging, injection, escaping, leaching, dumping, or disposing into the environment." Under EPCRA, the term "release" (§329(8)) includes emitting any hazardous chemical or extremely hazardous substance into the environment. Superfund excludes the "normal application of fertilizer" from the definition of release, and EPCRA excludes from the definition of hazardous chemicals any substance that is "used in routine agricultural operations or is a fertilizer held for sale by a retailer to the ultimate customer."

The CERCLA definition of "hazardous substance" (§101(14)) triggers reporting under both laws. Among the reportable substances that may be released by livestock facilities are hydrogen sulfide, ammonia, and phosphorus. The reportable quantity (RQ) for both hydrogen sulfide and ammonia is 100 pounds per day, or 18.3 tons per year; the RQ for phosphorus is 1 pound per day. Section 109 of Superfund and Section 325 of EPCRA authorize EPA to assess civil penalties for failure to report releases of hazardous substances that equal or exceed their reportable quantities (up to $27,500 per day under CERCLA and $27,500 per violation under EPCRA). Requirements of both can be enforced by citizens under provisions of the laws, which allow "any person" to commence a civil action against either a person who violates a legal prohibition or requirement, or against EPA for failure to perform a nondiscretionary duty or specified actions (CERCLA §310, EPCRA §326).

In addition to these reporting requirements, Superfund includes provisions authorizing federal cleanup of releases of hazardous substances, pollutants, or contaminants that may present an imminent and substantial danger to the public health or welfare (§104) and imposing strict liability for cleanup and damages to natural resources from releases of hazardous substances (§107). The applicability of these provisions to animal agricultural sources and activities has increasingly been receiving attention.

ENFORCEMENT AGAINST AFOS

EPA has enforced the Superfund and EPCRA reporting requirements against AFO release of hazardous pollutants in two separate cases. The first involved the nation's second-largest pork producer, Premium Standard Farms (PSF) and Continental Grain Company. In November 2001, EPA and the Department of Justice announced an agreement resolving numerous claims against PSF concerning principally the Clean Water Act, but also the Clean Air Act, Superfund, and EPCRA. More recently, in September 2006, the department announced settlement of claims against Seaboard Foods—a large pork producer with more than 200 farms in Oklahoma, Kansas, Texas, and Colorado—and PIC USA, the former owner and operator of several Oklahoma farms now operated by Seaboard. Like the earlier PSF case, the government had brought complaints for violations of several environmental laws, including failure to comply with the release reporting requirements of CERCLA and EPCRA.

The citizen suit provisions have been used to sue poultry producers and swine operations for violations of the laws. In two cases, environmental advocates claimed that AFO operators had failed to report ammonia emissions, putting them in violation of Superfund and EPCRA. In both cases, federal courts supported broad interpretation of key terms defining applicability of the laws' reporting requirements to livestock operations.

In the first of these cases, a federal district court in Oklahoma initially ruled in 2002 that a farm's individual barns, lagoons, and land application areas are separate "facilities" for purposes of CERCLA reporting requirements, rather than aggregating multiple emissions of pollutants across the entire site. This court held that Superfund's reporting requirements would only apply if emissions for *each individual facility* exceed 100 pounds per day. However, the district court's ruling was reversed on appeal (*Sierra Club v. Seaboard Farms Inc.,* 387 F.3d 1167 (10th Cir. 2004)). The court of appeals ruled that the whole farm site is the proper entity to be assessed for purposes of CERCLA reporting and determining if emissions of covered hazardous substances meet minimum thresholds.

In the second case, a federal district court in Kentucky similarly ruled in 2003 that the term "facility" should be interpreted broadly to include facilities operated together for a single purpose at one site, and that the whole farm site is the proper entity to be assessed for purposes of the Superfund and EPCRA reporting requirements (*Sierra Club v. Tyson Foods, Inc.,* 299 F. Supp. 2d 693 (W.D. Ky. 2003)). While Superfund provides that a continuous release is subject to reduced reporting requirements, and EPCRA provides an exemption for reporting releases when the covered substance is used in routine agricultural operations or is used on other farms for fertilizer, the court found that these exemptions did not apply to the facts of this case. The ruling was not appealed.

EPA was not a party in either of these lawsuits. The U.S. Court of Appeals for the 10th Circuit invited EPA to file an amicus brief in the *Seaboard Farms* case in order to clarify the government's position on the issues, but EPA declined to do so within the time frame specified by the court.

Three other cases in federal courts, while they do not include reporting violations, also have drawn attention, in part because they raised the question of whether animal wastes that contain phosphorus are hazardous substances that can create cleanup and natural resource injury liability under Superfund.[5] Animal wastes typically contain low levels of phosphorus, and animal wastes are beneficially used as fertilizer on farms. Over the long term, however, the application of animal waste fertilizers may result in phosphorus buildup in soils which may be released to watersheds through surface runoff. In 2003, a federal court in Oklahoma held that phosphorus contained in poultry litter in the form of phosphate is a hazardous substance under Superfund and thus could subject poultry litter releases to provisions of that law (*City of Tulsa v. Tyson Foods, Inc.*, 258 F. Supp. 2d 1263, (N.D. Okla. 2003)). This ruling was later vacated as part of a settlement agreement, but some observers believe that the court's reasoning may still be persuasive with other courts. The second case, *City of Waco v. Schouten* (W.D. Tex., No. W-04- CA-118, filed April 29, 2004), was brought by the city against 14 dairies alleging various causes of action based on disposal of wastes from those operations. It was resolved by a settlement agreement early in 2006. The third case, *State of Oklahoma v. Tyson Foods, Inc.* (N.D. Okla., No. 4:05-cv-00329, filed June 13, 2005), was brought by the Oklahoma Attorney General, asserting various claims based on the disposal of waste from 14 poultry operations in the Illinois River Watershed. The state principally sought response costs and natural resource injuries under CERCLA due to release of wastes from these facilities. The case was dismissed in 2010.[6]

The net result of these lawsuits is growing concern by the agriculture community that other legal actions will be brought and that the courts will continue to hold that the Superfund and EPCRA reporting requirements and other provisions apply to whole farm sites, thus potentially exposing more of these operations to enforcement under federal law.

Administrative Reporting Exemption

In 2005, a group of poultry producers petitioned EPA for an exemption from EPCRA and CERCLA emergency notification requirements for releases of ammonia, arguing that such releases from poultry growing operations pose little or no risk to public health, while reporting imposes an undue burden on the regulated community and government responders.[7] In 2007, EPA formed an internal workgroup to review information on animal waste as it relates to CERCLA and to possible exemptions from emissions reporting, and EPA Administrator Stephen Johnson told congressional committees that the agency would propose a rule to exempt routine animal waste air releases from emergency notification requirements.

In December 2007, EPA issued a proposal in response to the poultry industry petition.[8] It proposed to exempt releases of hazardous substances to the air (typically during digestion or decomposition) from animal waste at farms from the notification requirements of CERCLA and EPCRA. The exemption would apply to releases to the air from manure, digestive emissions, and urea, including animal waste mixed with bedding, compost, and other specified materials. "Farm" was defined in the proposal as an agricultural operation from which $1,000 or more of agricultural products are sold annually (the same definition used by the Department of Agriculture). EPA explained that the rule is justified because of the resource burden to industry of complying with reporting requirements, since the agency could not foresee a situation where a response action would be taken as a result of notification of releases of hazardous substances from animal waste at farms.[9]

The proposal drew significant public response during the comment period. While the proposed exemption pleased many agriculture industry groups who seek a waiver or other means to limit possible liability under CERCLA and EPCRA, environmental advocates and other stakeholders opposed the exemption, saying that emissions from animal wastes are not trivial or benign. Critics noted that the EPA proposal would exempt releases of ammonia, as originally requested in the industry petition, plus hydrogen sulfide and all other hazardous chemicals, such as nitrous oxide and volatile organic compounds released from animal waste. Some argued that an exemption was premature, since EPA was moving forward with research on emissions levels, which could be undermined by a regulatory exemption.[10] State air quality officials recommended that if the agency were to consider any action, it should only be a narrow exemption, such as one based on a size threshold for farms.[11]

In September 2008, the Government Accountability Office (GAO) issued a report evaluating EPA's activities to regulate air emissions and water discharges from animal feeding operations. GAO found that EPA is unable to assess the extent to which pollution from feedlots may be impairing human health and the environment, because it lacks data on the amount of pollutants that CAFOs are releasing to the air and water. GAO recommended that EPA develop a comprehensive national inventory of CWA-permitted CAFOs and accelerate its efforts to develop protocols for measuring and quantifying air contaminants from animal feedlots. GAO noted that EPA has been criticized because its current air emissions monitoring activities are limited in scope and sample size and may not produce sufficient information to shape future regulation. Moreover, GAO questioned the basis for the CERCLA/EPCRA exemption that EPA proposed in 2007. "It is unclear how EPA made this determination when it has not yet completed its data collection effort and does not yet know the extent to which animal feeding operations are emitting these pollutants."[12]

In December 2008, EPA finalized the CERCLA/EPCRA administrative reporting exemption with some modifications to the original proposal.[13] The final rule exempts hazardous substance releases that are emitted to the air from animal waste at farms from the notification requirement of CERCLA. Like the proposal, the final rule relieves *all* livestock operations of all size, not just poultry farms, from CERCLA's requirement to report hazardous substance releases to the air to federal officials. In addition, the final rule provides a partial exemption for such releases from EPCRA's requirement to report releases to state and local emergency officials. Partially responding to public comments, the final rule continues to apply EPCRA's reporting requirement to large animal feeding operations (those that are subject to permitting requirements under the Clean Water Act)[14], but exempts smaller facilities. A number of groups criticized the final rule, again raising concern about the toxicity of chemicals such as ammonia and hydrogen sulfide that are emitted from animal waste facilities and arguing that CERCLA and EPCRA do not authorize administrative exemptions for specific industries.

In 2009, a coalition of environmental advocates challenged the rule in federal court, as did the National Pork Producers Council (*Waterkeeper Alliance v. EPA*, D.C. Cir., Nos. 09-1017, 09-1104 (consolidated)). Environmental advocates continue to argue that the entire rule is deficient, while the pork producers group objects to the fact that the rule only partially exempted releases from EPCRA. Other agriculture industry groups, such as the National Chicken Council, intervened in the litigation in support of the final rule. Parties to the litigation entered into talks to mediate the issues, but in June 2010, the federal government asked to remand the final rule, without vacature, for EPA to reevaluate and possibly modify the rule. The court approved the government's request in October 2010. EPA does not have an announced schedule for proposing a new or modified rule. According to press reports, EPA does not plan to directly regulate the emissions, but is seeking to require their reporting. In the meantime, the 2008 exemption rule is in effect.

Under the exemption rule, large CAFOs remain subject to EPCRA's reporting requirements. Facilities' compliance with the rule has drawn attention: in July 2012, the Humane Society notified 51 swine facilities in Iowa, Oklahoma, and North Carolina that it intended to sue them over alleged unreported releases of ammonia.[15] As of September 2013, no lawsuits have been filed.

CONGRESSIONAL INTEREST

The court cases testing the applicability of Superfund and EPCRA to poultry and livestock operations have led to congressional interest in these issues. In March 2004, a number of Senators wrote to the EPA Administrator to ask the agency to clarify the reporting requirements of the two laws so as to limit their impact on poultry operations. The Senators' letter said that because of unclear regulations and a lack of scientific information about emissions, poultry and livestock producers are uncertain about the laws' requirements and are vulnerable to enforcement actions.[16] In report language accompanying EPA's FY2006 appropriations, the House Appropriations Committee urged EPA to address the issues.

> The Committee continues to be concerned that unclear regulations, conflicting court decisions, and inadequate scientific information are creating confusion about the extent to which reporting requirements in [CERCLA] and [EPCRA] cover emissions from poultry, dairy, or livestock operations. Producers want to meet their environmental obligations but need clarification from the Environmental Protection Agency on whether these laws apply to their operations. The committee believes that an expeditious resolution of this matter is warranted.[17]

Also in 2004, some in Congress considered proposing legislation that would amend the definition of "release" in Superfund (§101(22); 42 U.S.C. §9601(22)) to clarify that the reporting requirements do not apply to releases from biological processes in agricultural operations and to amend EPCRA to exclude releases of hazardous chemicals produced through biological processes in routine agricultural operations.

Some Members sought to amend the FY2006 Agriculture appropriations bill, H.R. 2744, with a provision exempting releases of livestock manure from CERCLA and EPCRA. The proposal was promoted by Senate conferees on the bill, but it was not accepted by House conferees. Proponents, including

Senator Larry Craig, contended that the proposed language was consistent with current law, because in their view CERCLA and EPCRA were never intended to apply to agriculture. Environmentalists objected to the language, arguing that it could prevent public health authorities from responding to hazardous substance releases from AFOs, would block citizen suits against agriculture companies for violations of reporting requirements, and would create an exemption from Superfund liability for natural resource injuries that might result from a large manure spill. EPA's congressional affairs office released an unofficial analysis criticizing the bill. It argued that, by eliminating federal liability for manure releases under Superfund and EPCRA, the provision could interfere with EPA's Air Compliance Agreement, because companies would have much less incentive to participate in the agreement. The agreement is a plan that EPA announced in January 2005 to collect air quality monitoring data on animal agriculture emissions.[18] The House and Senate gave final approval to H.R. 2744 in November 2005 (P.L. 109-97), without the language that Senate conferees had proposed.

Also in November 2005, legislation was introduced in the 109th Congress to amend CERCLA to clarify that manure is not a hazardous substance, pollutant, or contaminant under that act and that CERCLA's notification requirements would not apply to releases of manure (H.R. 4341). The bill was similar to the legislative language that Senator Craig had proposed to conferees as a provision of the FY2006 Agriculture appropriations bill with a broad definition of "manure" that includes, for example, bedding commingled with animal waste.

H.R. 4341 was introduced the same day that a House Energy and Commerce subcommittee held a hearing on animal agriculture and Superfund. The Subcommittee on Environment and Hazardous Materials heard from agriculture industry witnesses who urged Congress to provide policy direction on the issue that has developed as a result of recent and potential litigation. Other witnesses testified that the reporting and notification requirements of Superfund and EPCRA provide a safety net for making information on releases available to government and citizens, and that other environmental laws, such as the Clean Air Act, cannot function in that manner. An EPA witness said that the agency was at the time considering ways to reduce the paperwork burdens for large AFOs to report their emissions (resulting in the December 2007 proposed regulatory exemption, discussed above). Related legislation was introduced in the Senate (S. 3681). Similar legislation was introduced in the 110th Congress (H.R. 1398 and S. 807), but no further action occurred on any of these bills.

During consideration of farm bill legislation in the 110th Congress, the House Agriculture Committee approved an amendment expressing a sense of the committee that farm manure is not to be considered a toxic waste. However, the amendment was not included in the reported version of the bill, reportedly because of jurisdictional issues, and was not included in the final legislation enacted in 2008 (P.L. 110-246).

Some Members of Congress were critical of EPA's 2007 proposal to exempt routine animal waste air releases from CERCLA and EPCRA's reporting requirements, questioning the potential for harmful environmental and enforcement impacts of the proposal.[19] At a September 2008 House Energy and Commerce subcommittee hearing where GAO's report was discussed (see page 5), several Members said that they were skeptical of EPA's authority to authorize a blanket exemption. Others suggested that an exemption for small farms, whose emissions are unlikely to cause environmental harm, would make sense. EPA and USDA witnesses supported the proposed rule, saying that the air release waiver would only affect reporting meant for emergency response situations, but would not affect requirements to report emissions of hazardous substances from other farm sources, or releases of hazardous substances from manure into soil, ground water, or surface water.[20]

Legislation on this topic was introduced in the 112th Congress. Companion bills, H.R. 2997 and S. 1729, were similar to bills introduced in earlier Congresses. Both were intended to clarify that manure is not a "hazardous substance" or "pollutant or contaminant" under CERCLA and to remove emissions reporting liability under CERCLA and EPCRA. Supporters of these bills seek to block EPA from revising the 2008 exemption rule so as to require reporting of releases. A House Energy and Commerce subcommittee held a hearing on H.R. 2997 on June 27, 2012. At that hearing, an EPA witness said that the agency "has concerns with the broad impacts of H.R. 2997," which the witness described:

> The effect of the bill would be to prevent the EPA from using CERCLA response authorities to respond to releases to the environment when manure is the source of those hazardous substances, even if the release, for instances such as the failure of a large manure waste lagoon, presented a substantial danger to the public health and the environment. It would also prevent the Agency from issuing CERCLA abatement orders to require response to damaging releases.[21]

In the 113th Congress, no similar legislation has been introduced.

POLICY ISSUES

Supporters and opponents of legislation since the 109th Congress have raised a number of arguments for and against a statutory exemption from CERCLA and EPCRA. For example, proponents of the exemption proposed in these bills, representing the agriculture industry, especially livestock and poultry producers, said that animal manure has been safely used as a fertilizer and soil amendment by many cultures all over the world for centuries and thus should not be considered a hazardous substance. Opponents—including environmental activists, public health advocates, and state and local governments—agreed that when properly managed, manure has beneficial uses. Superfund's reporting and cost recovery requirements do not threaten responsible operators who manage manure as a valuable fertilizer, they said. However, these groups said that when improperly managed and in the massive amounts produced at today's large feedlot operations, animal waste can release a number of polluting substances to the environment. Releases to surface water, groundwater, and the atmosphere may include nutrients, organic matter, solids, pathogens, volatile compounds, particulate matter, antibiotics, pesticides, hormones, gases that are associated with climate change (carbon dioxide and methane), and odor.

Proponents of the legislation argued that neither Superfund nor EPCRA was intended by Congress to apply to agriculture and that the pending legislation would simply clarify congressional intent. CERCLA exempts "normal application of fertilizer" from the definition of "release" and also exempts releases of "naturally occurring organic substances." Animal waste arguably was intended to be covered by these existing exemptions, they said. Opponents responded that there is little firm evidence either way on this point, as there is limited legislative history concerning this language. The exemption for "normal application of fertilizer," enacted in CERCLA in 1980, applies to application of fertilizer on crops or cropland for beneficial use, but does not mean dumping or disposal of larger amounts or concentrations than are beneficial to crops.[22]

EPA has not issued guidance to interpret what constitutes "normal application of fertilizer," and the only court decision so far addressing this issue (the vacated 2003 *City of Tulsa* case discussed above) held that neither plaintiffs nor defendants in that case had presented evidence sufficient for a fact-based determination of what constitutes "normal application." Opponents of the legislation also argued that animal manure consists of a number of substances that are nutritional and pharmaceutical elements of the feed provided to animals (trace elements, antibiotics, nutrients), and releases are the result of inadequate waste disposal, not "naturally occurring" substances and activities.

Proponents argued that enforcement and regulatory mechanisms exist under the Clean Water Act (CWA) and other media-specific statutes, such as the Clean Air Act (CAA), making it unnecessary to rely on Superfund or EPCRA for enforcement or remediation. In particular, both the Clean Water Act and Clean Air Act require that regulated facilities obtain permits that authorize discharges or emissions of pollutants. Enforcement of permit requirements has been an important tool for government and citizens to address environmental concerns of animal agriculture activities.

Opponents responded that enforcement under Superfund fills critical gaps in these other environmental laws, because not all pollutants are covered by other laws. For example, releases of ammonia and hydrogen sulfide are listed under CERCLA but are not currently regulated as hazardous pollutants under the CAA. Clean Water Act AFO permits primarily address discharges of nutrients, but not other components of manure waste (e.g., trace elements, metals, pesticides, pathogens). Moreover, neither of these laws provides for recovery of costs for responding to or remediating releases, nor for natural resource injuries. Opponents also argued that, while "federally permitted releases" are exempt from CERCLA's reporting requirements, CWA and CAA permit requirements apply only to facilities that meet specified regulatory thresholds (for example, CWA permit rules apply to about 15,300 large AFOs, less than 6% of all AFOs in the United States).

Finally, proponents of the legislation have argued that if animal manure is considered to be a hazardous substance under Superfund, farm operations both large and small potentially could be exposed to costly liabilities and penalties. Opponents have noted that the purpose of release reporting is to keep federal, state, and local entities informed and to alert appropriate first responders of emergencies that might necessitate response, such as release of hazardous chemicals that could endanger public health in a community. The exemption proposed in legislation, they pointed out, would apply not only to CERCLA and EPCRA reporting requirements but also to other provisions (such as Superfund's authority for federal cleanup of releases, cleanup liability, and liability for natural resource injuries).

According to states and some other interest groups, liability, which arises when manure is applied in amounts that exceed what is beneficial to support crops, is necessary to bring about improvements in waste handling practices of large AFOs. Enacting an exemption would severely hamper the ability of government to appropriately respond to releases of hazardous substances and pollution caused by an animal agriculture operation, they argued. On the issue of penalties, opponents noted that penalties are not available under Superfund for removal or remedial actions (except for failure to comply with information gathering and access related to a response action), regardless of whether initiated by government or a private party. CERCLA does authorize civil penalties for violation of the Section 103 reporting requirements (up to $37,500 per day), but neither of the two key citizen suit cases decided thus far (*Sierra Club v. Tyson Foods, Inc.*, and *Sierra Club v. Seaboard Farms Inc.*) involved penalties for failure to report releases.

CONCLUSION

Issues concerning the applicability of Superfund and EPCRA to animal agriculture activities have been controversial and have drawn considerable attention. Bills introduced in Congress on several occasions to clarify that manure is not a "hazardous substance" or "pollutant or contaminant" under CERCLA and to remove emissions reporting liability under CERCLA and EPCRA gained support, but none was enacted. They also drew opposition from environmental advocacy groups and state and local governments. The Bush Administration did not present an official position on the legislation. The Obama Administration's views on the issue are reflected in EPA's anticipated reconsideration of the 2008 rule that exempts air releases of hazardous substances from animal waste at farms from CERCLA and most EPCRA reporting, and in opposition to H.R. 2997 in the 112th Congress.

End Notes

[1] For additional information, see CRS Report RL33325, *Livestock Marketing and Competition Issues*, by Renée Johnson and Geoffrey S. Becker.

[2] U.S. Department of Agriculture, Natural Resources Conservation Service, "Manure Nutrients Relative to the Capacity of Cropland and Pastureland to Assimilate Nutrients: Spatial and Temporal Trends for the United States," Publication no. nps00-579, December 2000, p. 85.

[3] For additional information, see CRS Report RL32948, *Air Quality Issues and Animal Agriculture: A Primer*, by Claudia Copeland.

[4] For additional information on Superfund and EPCRA, see CRS Report RL30798, *Environmental Laws: Summaries of Major Statutes Administered by the Environmental Protection Agency*, coordinated by David M. Bearden, and CRS Report RL33426, *Superfund: Implementation and Selected Issues*, by Jonathan L. Ramseur.

[5] Unlike the citizen suit cases discussed above, these lawsuits do not address what is a "facility," for purposes of determining whether a release has occurred. EPA also was not a party in any of these cases.

[6] Environmental advocacy groups also have brought lawsuits against agricultural operations under the Resource Conservation and Recovery Act (RCRA), the federal law that governs the treatment, storage, and disposal of solid and hazardous waste. For example, in February 2013 environmental groups challenged the waste management practices of several dairies in Washington state. The lawsuits do not allege that manure is hazardous waste. Rather, they allege that the dairies over-applied and improperly applied manure to their fields and allowed liquid manure to leak from lagoons, thus qualifying as "solid waste" under RCRA. They seek to require the dairies to upgrade their manure storage facilities. See, for example, Community Association for Restoration of the Environment and Center for Food Safety v. Cow Palace, LLC (E.D. WA, No. 13-CV-3016-TOR, filed February 14, 2013).

[7] In 1998, EPA granted an administrative exemption from release reporting requirements for certain radionuclide releases. EPA cited authority in CERCLA sections 102(a), 103, and 115 for granting administrative reporting exemptions where "releases of hazardous substances that pose little or no risk or to which a Federal response is infeasible or inappropriate." See 63 *Federal Register* 13461 (March 19, 1998).

[8] U.S. Environmental Protection Agency, "CERCLA/EPCRA Administration Reporting Exemption for Air Releases of Hazardous Substances from Animal Waste," 72 Federal Register 73700 (December 28, 2007).

[9] Ibid., p. 73704.

[10] See CRS Report RL32947, *Air Quality Issues and Animal Agriculture: EPA's Air Compliance Agreement*, by Claudia Copeland.

[11] National Association of Clean Air Agencies, letter to the Honorable Barbara Boxer, chairman, Senate Environment and Public Works Committee, March 20, 2007.

[12] U.S. Government Accountability Office, "Concentrated Animal Feeding Operations, EPA Needs More Information and a Clearly Defined Strategy to Protect Air and Water Quality from Pollutants of Concern," September 2008, GAO- 08-944, p. 7.

[13] U.S. Environmental Protection Agency, "CERCLA/EPCRA Administrative Reporting Exemption for Air Releases," 73 *Federal Register* 76948-76960, December 18, 2008.

[14] For additional information, see CRS Report RL31851, *Animal Waste and Water Quality: EPA Regulation of Concentrated Animal Feeding Operations (CAFOs),* by Claudia Copeland.

[15] EPCRA and other environmental laws that include citizen suit provisions require that citizen suitors must send a notice of intent to sue to appropriate persons at least 60 days before commencing an action. (EPCRA §326(d), 42 U.S.C. §11046(d)) See http://www.humanesociety.org/news/press_releases/ 2012/07/ hsus_intent_to_sue_pig_ confinement_facilities_071112.html.

[16] Senator Blanche L. Lincoln et al., letter to Michael Leavitt, EPA Administrator, March 12, 2004.

[17] U.S. Congress, House Committee on Appropriations, *Report accompanying H.R. 2361, Department of the Interior, Environment, and Related Agencies Appropriation Bill*, 2006, H.Rept. 109-80, 109th Cong., 1st sess., p. 87.

[18] For information, see CRS Report RL32947, *Air Quality Issues and Animal Agriculture: EPA's Air Compliance Agreement*, by Claudia Copeland.

[19] Letter from Reps. John Dingell, Albert Wynn, Hilda Solis to Stephen L. Johnson, EPA Administrator, March 18, 2008.

[20] U.S. Congress, House Committee on Energy and Commerce, Subcommittee on Environment and *Hazardous Materials, Hazardous Substance Releases and Reporting under the Comprehensive Environmental Response, Compensation, and Liability Act of 1980 (CERCLA) and the Emergency Planning and Community Right-to-Know Act (EPCRA)*, 110th Cong., 2nd sess., September 28, 2008, Serial No. 110-151 (Washington: GPO, 2008).

[21] Testimony of Mathy Stanislaus, Assistant Administrator, Office of Solid Waste and Emergency Response, U.S. EPA, before the Subcommittee on Environment and the Economy, House Committee on Energy and Commerce, June 27, 2012.

[22] U.S. Senate, Committee on Environment and Public Works, *Environmental Emergency Response Act, Report to Accompany S. 1480*, 96th Cong., 2nd sess., S.Rept. 96-848, p. 46.

INDEX

D

E

F

G

H

I

U

V

W